Augsburger Schriften zur Mathematik, Physik und Informatik

Band 33

Edited by:
Professor Dr. B. Schmidt
Professor Dr. B. Aulbach
Professor Dr. F. Pukelsheim
Professor Dr. W. Reif
Professor Dr. D. Vollhardt

Bibliographic information published by the Deutsche Nationalbibliothek

The Deutsche Nationalbibliothek lists this publication in the Deutsche Nationalbibliografie; detailed bibliographic data are available in the Internet at http://dnb.d-nb.de .

ISBN 978-3-8325-4478-2
ISSN 1611-4256

Logos Verlag Berlin GmbH
Comeniushof, Gubener Str. 47,
10243 Berlin
Tel.: +49 030 42 85 10 90
Fax: +49 030 42 85 10 92
INTERNET: http://www.logos-verlag.de

Commutability of Γ-limits in problems with multiple scales

Dissertation

zur Erlangung des akademischen Grades

Dr. rer. nat.

eingereicht an der

Mathematisch-Naturwissenschaftlich-Technischen Fakultät

der Universität Augsburg

von

Martin Jesenko

Augsburg, Dezember 2016

Universität
Augsburg
University

Erstgutachter: Prof. Dr. Bernd Schmidt (Universität Augsburg)
Zweitgutachterin: Prof. Dr. Lisa Beck (Universität Augsburg)
Drittgutachter: Prof. Dr. Filip Rindler (University of Warwick)

Datum der mündlichen Prüfung: 10. April 2017

Preface

"Il matematico può trarre ispirazione dalle fonti più diverse, dalla fisica, dall'ingegneria, dall'arte, dall'economia, dal diritto, dalla filosofia. Non c'è forma di sapere da cui il matematico non possa trarre ispirazione. Anche fenomeni abbastanza semplici e ormai noti da parecchio tempo possono essere fonte di ispirazione di lavori originali per il matematico."

Ennio De Giorgi

It is a major challenge for mathematicians to describe real-world processes concisely in a mathematical form, to tackle them by applicable tools or devise appropriate new ones, and to find solutions that can be employed in actual problems. An important guideline is keeping a fine balance between the best possible approximation on one side and simplicity and effectiveness on the other. Furthermore, it should be taken into consideration that input data have limited accuracy so that a robust dependence on initial values is desirable.

This work draws the motivating ideas from material science. Although it is impossible to expect perfect regularity in real materials, it is in nature's nature to achieve balance which yields quite regular forms and patterns, at least at some scale and precision. On the other hand, empirical and theoretical studies have led to the construction of artificial materials with improved properties for intended purposes. A suitable treatment and mixture can lead to the required stiffness, conductivity, workability, to name just a few. The goal is a material with most possible regularity both for production and for quality reasons.

Having either a man-made or a natural composition, we are dealing with a material whose parameters differ from point to point, often on a considerably small (microscopic) scale compared to its dimensions. Theoretical materials that have all their properties equal in every point are called homogeneous (from Greek $o\mu o\gamma \varepsilon \nu \eta \varsigma$, of same kind). It is favourable to find for a given material such a homogeneous model that includes all arising issues and provides sufficiently accurate solutions on the normal (macroscopic) scale. Thus we would obtain a significantly simpler problem to cope with, not to speak of the demand of consumers to get the product data in an easily understood way. E.g., for an alloy, engineers require its Young's modulus and Poisson's ratio, yield and ultimate strength, thermal parameters, etc. All these characteristic values, which can be find in tables, effectively describe the macroscopic behaviour.

A good homogeneous model needs to provide results close to the actual ones. The strategy of the calculus of variations is to formulate the problem in an integral form with possible constraints and to determine the solutions as its extremal points. The important question is: how to recognize that one integral formulation is a good approximation of another, or, more precisely, a good limit of a sequence of such formulations. The key feature should be that the related extrema stay close to each other. Since its introduction by De

Giorgi, an extensively used tool at least for global extrema has been Γ-convergence. It ensures precisely this behaviour under some additional assumptions that yield compactness, see Theorem (B.2).

Clearly, the search for simpler models does not stop at the homogenization. Even more basic is the idea of a local approximation by a linear mapping. Depending on the loads, we might presume deformations of the material to be small and thus consider neglecting higher order terms. In this case, one speaks of geometric linearization to emphasize that the related problem need not be linear.

The behaviour of materials under mechanical loads can, roughly speaking, be divided into two parts. Until some point the deformations are reversible: if the material is released, it will take its original shape and structure. This is the elastic region which ends with the yield surface. In the linear elasticity, the relation between the stress and the strain is supposed to be linear, given by Hooke's law with the compliance tensor (or its inverse, the elasticity tensor) as the tensor of proportionality. Going beyond the yield surface results in permanent changes. The material behaves plastically and does not return to the reference configuration if unloaded. So-called hysteresis effects occur, and should such a material be reloaded, the new properties have to be known.

The problems in this work have their origins in both behaviours that, however, turn out to be mathematically considerably distinct. For that reason, this work is divided into two parts: the first is devoted to the study of problems related to elasticity models while the second explores a certain simplified elastoplasticity model.

As announced, the first part draws motivation from problems in elasticity theory. We look especially into two processes: homogenization and (geometric) linearization. An effective homogeneous theory provided by a homogenization procedure for a periodic stored-energy function of the material dates back to the works [Br:85] by Braides and [Mü:87] by Müller. On the other hand, however natural the concept of neglecting the higher order terms in the Taylor expansion might seem, the mathematically rigorous derivation of the linear elasticity from the nonlinear setting by means of Γ-convergence is non-trivial and has only been solved comparatively recently by Dal Maso, Negri and Percivale in [DNP:02]. They examine hyperelastic materials whose stored-energy function has a special, "one-well" structure with a single minimum and reasonable growth conditions. In this case the geometric linearization yields a simple linear problem.

If we have a material with a very fine periodic structure, and if the loads are small, then by the discussion above we have two possible effective models. This results in the following series of questions: Which one should be chosen? May we carry out both of them successively, and if so, in which order? Does the result depend on the order? Can we perform them simultaneously, and must we pay attention to the relation of scales? These questions were resolved by Müller and Neukamm in [MN:11] where they show that for materials with a periodic structure and a one-well stored-energy function the two processes are indeed interchangeable. The examination of simultaneous limits may be found in [Ne:10].

The linearization result from [DNP:02] was generalized by Schmidt in [Sch:08] to multiwell stored-energy functions which arise, e.g., in shape-memory alloys covered in [Bh:06]. Our starting question is whether also this (now only geometric) linearization commutes with homogenization. We start Part I with a motivational chapter where we present the mentioned results.

Instead of assuming periodicity, Braides in his work [Br:86] introduced homogenizable functions as the functions that allow for a homogeneous approximation in terms of Γ-convergence. He explored the question of "homogenization closure", i.e., in which sense must a homogenizable family be close to a single function in order for the latter to inherit this property. In Chapter 2 we choose an even more general approach of "Γ-closure". As the name suggests, we intend to determine under which proximity notion does arbitrary Γ-convergence, not necessary homogenization, transfer. This investigation is the core of Part I and was already the main topic of our joint article with Schmidt [JS:14]. We show that a sort of local uniform convergence, similar as in the works of Braides, is a sufficient assumption. First, we show the result for functionals with a standard growth. Since such a growth condition is too restrictive for applications in the elasticity theory, we relax in Section 2.3 the lower-bound inequality to a sort of Gårding inequality. Moreover, we show that boundary values may be imposed. Then we look at the starting problem of interchangeability of two Γ-limits. Under a stricter equivalence condition, we show in Section 2.6 that it proves to be the case. Finally, we apply the derived results to homogenization. We show that also for homogenizable, not necessarily periodic, functions that induce Gårding type functionals the standard multicell formula for the homogenized function holds true. The homogenization closure result is now merely a special case of the Γ-closure result from Section 2.3. Let us in the context of homogenization of multiparameter integrals mention the related works [AM:02, AM:04] of Alvarez and Mandallena.

The interest in such Γ-closure theory is twofold. Firstly, it sheds a new light to a number of problems that have been studied over the last years. In particular, we will see that the results on linearization in [DNP:02], on geometric linearization in [Sch:08] and on commutability in [MN:11, GN:11] can be to some extend seen as consequences of our abstract theorem. Secondly, our general method also allows for new applications. In Chapter 3 we show that it yields a positive answer to the above-stated question of the commutability in the case of multiwell energies. However, while all these examples concern the behaviour of elastic materials at small displacements, our general scheme is not restricted to applications in elasticity theory. In contrast, we presume that our results are of interest in a variety of different problems. Our Γ-closure theorems allow to analyse the effective properties of multiscale problems with two limiting parameters: They provide general criteria that guarantee commutability of the limits and stability results for simultaneous limits of the parameters. We believe that such results may be of particular interest, e.g., in numerical schemes for multiscale problems, where a thorough understanding of the interplay of the small model parameter and the length scale of the numerical discretization is crucial.

Chapter 3 is, as already mentioned, devoted to applications in elasticity. The geometric rigidity result of Friesecke, James and Müller in [FJM:02] implies that stored-energy functions in realistic models induce integral functionals of Gårding type. We thus see that indeed homogenization and geometric linearization commute. We formulate Theorem (3.9) in terms of general homogenizable densities and note that this includes, in particular, the case of periodic material mixtures. Moreover, we remark that all Γ-convergence statements can be complemented by observing compactness for finite energy sequences with pre-assigned boundary values, see Remark (3.12). We show that the results from Chapter 2 imply also the linearization results in [DNP:02] and [Sch:08].

The discussions about homogenization were made under a periodicity assumption, though we allowed for other densities with a similar behaviour. In other words, the structure of the material must be really perfect, without any errors, in order for the results to hold without reservations. This may be loosened to some extend by relocating the discussion from the deterministic to a stochastic environment. Although now integral functionals appear only with a given probability, if the distribution is well-behaved, i.e. periodic in law and ergodic, we still get a deterministic homogeneous effective model. The fundament of the ergodic theory represents the Akcoglu-Krengel ergodic theorem in [AK:81], which was generalized in [LM:02]. The investigation of stochastic homogenization began with a couple of works from Dal Maso and Modica [DM:86-1, DM:86-2], where they consider densities that are convex in the second variable. A stochastic counterpart of the homogenization theory under standard growth conditions without the convexity assumption was proved in [MM:94]. In Chapter 4 we, analogously to the deterministic setting, extend this result to random functionals of Gårding type and give stochastic versions also of other results from Chapter 2. Let us mention that the stochastic analogue of the commutability result from [MN:11] for one-well densities was proved already in [GN:11]. Duerinckx and Gloria wrote recently a nice survey [DG:16] on the development of both deterministic and stochastic homogenization with improved results. Some recent results on the stochastic homogenization may be found also in [Ze:16].

After all the considerations for the p-growth with $p > 1$, we ask in Chapter 5 the obvious question if the same holds for $p = 1$. Not surprisingly, the answer is negative due to the lack of equiintegrability of gradients of bounded sequences in $W^{1,1}$ that distinguishes this setting from the others. We will borrow the idea for a concrete counterexample from [BD:93] and provide an additional equivalence notion that allows for Γ-closure results to hold also for $p = 1$. However, since our interest in this topic is purely for the sake of completeness, we will show only the most basic Γ-closure result. Should there be a demand for any related subsequent result, it is possible to adapt the analysis from the previous chapters, though with additional equivalence assumptions.

In Part II we turn our attention to the models from elastoplasticity. A mathematical description of the plastic behaviour with different models can be found, e.g., in the book by Duvaut and J. Lions [DL:76] as well as in the newer one by Han and Reddy [HR:99]. As previously mentioned, after surpassing the yield point, the material undergoes plastic deformations and does by unloading not return to the original state. Therefore, for a thorough analysis of its behaviour, the knowledge of its history is necessary, which makes the problem time-dependent, i.e. really dynamic or at least quasistatic. This setting was covered by several authors starting with Suquet in [Su:81]. A contemporary result on the quasistatic approach with some references can be found in [Mo:16]. Nonetheless, there is also the corresponding static theory, known as the Hencky plasticity. Clearly, its validity is restricted to one-time loading since it cannot include hysteresis effects.

The simplest model of plastic behaviour is the perfect plasticity. It is said that a material behaves perfectly plastic if in the plastic region deformations cost no energy, i.e. there is no rise of stress. We will also use the term "zero hardening" as opposed to the models with linear hardening that will be also the subject of our considerations. The boundary between the elastic and the plastic region, the yield surface, is given by yield criteria. During the past, several of these criteria were proposed, and the most commonly used one (at least for ductile materials) is the von Mises criterion. It bounds the shear energy density and thus leaves the trace-part of the stress unconstrained. This results in a special growth property of the stored-energy function: it is linear in the deviatoric part of strain and quadratic in the trace part. Obviously, the results from Part I are not applicable.

The model with the perfectly plastic behaviour and the von Mises yield criterion was investigated in [AG:82, Te:85, BDV:97], however for homogeneous materials. The stored-energy function on indeed has special growth properties but it splits into two summands: the linear is given by a convex function of the deviatoric part of strain, and the quadratic is simply a multiple of the squared trace. Our goal is to generalize the structure of the stored-energy function and assume solely the typical "Hencky plasticity" growth. Moreover, we intend to allow for location-dependence and explore potential homogenizability for periodic densities. A similar setting though for convex densities was considered in [DQ:90]. In contrast to Part I, here we do not seek a fully general theory but restrict our considerations closely related to the Hencky plasticity model, the advantage being that the appropriate spaces of functions have already been introduced and investigated. We believe, however, that some steps towards the result can be adapted for other problems with different growth properties in different directions. Or at least, one could check which of the essentially employed tools in our analysis are still available.

In Chapter 6 we give a very brief introduction of the plasticity models that we consider. Then we set and present the problem. Once more we have two processes, one again being homogenization and the other vanishing hardening in the plastic region, and we wish to find out whether they are interchangeable. The vanishing hardening is fairly easy to treat as we are dealing with a decreasing sequence. The main task is, therefore, to explore the potential homogenizability of periodic functions with the growth typical for Hencky plasticity. We conclude the chapter by investigating a related one-dimensional problem that might be looked upon as an extension of the one-dimensional quadratic case.

The preliminary work is included in Chapter 7. The structure of the energy functional requires the introduction of appropriate spaces of functions. Therefore, we introduce the space of functions of bounded deformation and its relevant subspaces with different topologies. Two results are crucial for the solving our problem. Firstly, even functions of bounded deformation are approximately differentiable in the same meaning as are BV-functions, and it is possible to upgrade this property to a higher order L^q-differentiability. Secondly, we essentially use the fact that the singular part of the symmetrized gradient Eu has a (symmetric) rank-one structure. This analogue to Alberti's rank one theorem for BD-functions was proved recently by De Philippis and Rindler in [DR:16].

In Chapter 8, after recalling the setting, we first introduce the natural candidate for the homogenized density, having in mind the results in Part I and general results for subadditive processes from [LM:02]. The upper bound for the Γ-lim sup follows from a continuity result in the $\langle\cdot\rangle$-strict topology. The proof employs the Reshetnyak continuity theorem and the rank one theorem for BD, and it draws ideas from [KR:10-1]. However, there is quite some technical work required due to the quadratic growth of the trace part. We assess this part by a sort of Lipschitz inequality that we deduce from the local Lipschitz continuity result in [BKK:00]. Regarding the lim inf-inequality, we may adapt the slicing method of De Giorgi for the regular points by the use of Bogovskii's operator and the L^q-differentiability of functions of bounded deformation. Thus, we get a homogenization result at least on the space LU. In order to control the singular points, we will have to suppose an asymptotically convex behaviour so as to use the results from [DQ:90]. With this additional assumption we get a full homogenization result for periodic functions with Hencky plasticity growth, and we can also show that homogenization and vanishing hardening commute. In Section 8.6 we investigate the related relaxation problem without the asymptotic convexity condition but with an additional growth assumption. Thus we may apply new results on convexity from Kirchheim and Kristensen in [KK:16].

On the subject of prerequisites, Sobolev spaces are assumed to be a known field. Therefore, there are no definitions, theorems or even references regarding them. The same holds also for the space of functions of bounded variation that play a minor role anyway. On the other hand, we consider the space of functions of bounded deformation to be quite specialized, and we therefore give all necessary definitions and properties, at least by references. Clearly, the same applies to its subspaces specially designed for the problems in the Hencky plasticity. In Appendix we state some applied tools, which might assist the reader. We include different types of convexity, mostly for its symmetric counterparts that appear comparatively seldom in the literature. Since it is the main topic of the work, there is also a short survey on the Γ-convergence.

This preface should serve as a general invitation and as an orientation. Its purpose was to introduce the topic, explain the origins and give a general overview of the scope of the work. Each chapter contains at the beginning a more detailed preview of the aims, the applied tools and our accomplishments.

Acknowledgements

"Then how does each of us hear them in his own native language?" [Acts 2:8]

Meine erste mathematische Liebe war und bleibt die Analysis. Prof. Bernd Schmidt hat mir die Möglichkeit gegeben, in diesem Bereich zu promovieren, was an meiner ehemaligen Universität nicht möglich wäre. Dafür und für das Vertrauen, die Geduld und alle investierte Zeit bei der Betreuung möchte ich mich bei ihm herzlich bedanken.

Es ist nicht selbstverständlich, wenn man sich auf seiner Arbeitsstelle wohlfühlt. Deswegen bin ich allen Kolleg(inn)en für die gute Stimmung am Lehrstuhl für Nichtlineare Analysis sehr dankbar. Mit Julian Braun haben wir in unserem Büro ein stilles Arbeitsumfeld geschaffen, dass viel Neid erweckt hat. Die Gespräche über Fußball beim Kakao habe ich sehr genossen, obwohl meine Versuche, anderen etwas beizubringen, allzu oft gescheitert sind. Und mit Veronika konnte ich die Leidenschaft für die BD-Funktionen teilen.

Naturalmente vorrei ringraziare anche i nostri tre italiani. ¡Y a los castellanohablantes también!

Najlepše pa bi se rad zahvalil svojim staršem ter sestrama Ani in Tamari, ki so verjetno bolj kot jaz verjeli, da mi bo uspelo, in mi tudi zato ali pa prav zato tudi je. Nekega dne bom izvedel, kdo vse je pridodal svoj kamenček v mozaik in kakšna skala je to bila. V veliko zadovoljstvo mi je dejstvo, da sta priprava in zagovor tega dela potekala v času jubilejnega leta mojega zavetnika. Brez Marijinega varstva in vodstva pa bi bil tako ali tako vsak trud zaman.

Table of contents

Part I

General Γ-closure and commutability results with applications to homogenization and elasticity theory

Chapter 1

Motivation from elasticity

This chapter presents some concepts and results in elasticity theory that will serve as a motivation for a further study. We present the process of geometric linearization for the one-well case, which justifies the standard theory of linear elasticity, as well as for the multiwell case. Moreover, we explain the derivation of homogeneous models. In Section 1.5 we pose the ensuing question of relation between the two processes. We state Theorem (1.1) whose proof was the initial goal of our considerations.

1.1 Variational formulation

Elastic behaviour of some material can be described in the energetic form. If the material is hyperelastic, then the elastic energy has a density function. More precisely, let a bounded Lipschitz domain $\Omega \subset \mathbb{R}^n$ represent the reference configuration of some hyperelastic body. Then there exists the stored-energy function $W : \Omega \times \mathbb{R}^{n \times n} \to \mathbb{R}$ such that for any admissible deformation $y : \Omega \to \mathbb{R}^n$ the corresponding elastic energy is

$$\mathcal{E}(y) = \int_\Omega W(x, \nabla y(x))\ dx.$$

The stored-energy function is expected to be *frame-indifferent*, i.e.,

$$W(x, RX) = W(x, X) \quad \text{for all } x \in \Omega,\ X \in \mathbb{R}^{n \times n} \text{ and } R \in \mathrm{SO}(n).$$

Therefore, as we will see shortly, W depends in the second argument in the physically reasonable states only on $X^T X$.

Minimizers of the total energy, which may also contain potential energies of external forces, represent equilibrium states. We will restrict our discussions to the global minimizers.

When dealing with concrete problems, we expect certain properties of the solutions. We try to incorporate this into the model, which may lead to a simpler problem. However, we must justify this step by showing that the corresponding minimizers stay close. The right tool for such limiting processes is the concept of Γ-convergence of functionals (see Appendix B).

1.2   Linear elasticity

If the reference configuration is stress-free, and the loads are small, we also expect displacements to be small. Hence, considering only the linear terms and neglecting the higher order ones should yield a good approximation of the original problem. Since the new problem will be geometrically linear, and therefore considerably easier, we speak of linear elasticity. Let us explain the setting more in detail. For the complete description with the proofs of the results, see [DNP:02].

We suppose that the stored-energy function $W : \Omega \times \mathbb{R}^{n \times n} \to \mathbb{R}$ has the only minimum, say 0, at the reference configuration. Because of frame-indifference, for every $x \in \Omega$

$$W(x, X) = 0 \iff X \in \mathrm{SO}(n).$$

It should grow at least quadratically away from zero set, i.e., for some $C > 0$

$$W(x, X) \geq C \operatorname{dist}^2(X, \mathrm{SO}(n))$$

for all $x \in \Omega$ and $X \in \mathbb{R}^{n \times n}$.

Since the displacements are small, we introduce a small parameter $\delta \ll 1$ and rescale the deformation and the stored-energy function as

$$y(x) = x + \delta u(x) \quad \text{and} \quad V_\delta(x, Y) := \frac{1}{\delta^2} W(x, I + \delta Y).$$

Physically admissible deformations are orientation-preserving, i.e. $\det \nabla y(x) > 0$. The polar decomposition of such matrices has the form

$$X = R\sqrt{X^T X}$$

with $R \in \mathrm{SO}(n)$, and by the frame indifference

$$W(x, X) = W(x, \sqrt{X^T X}).$$

In our case is

$$(\nabla y(x))^T (\nabla y(x)) = (I + \delta \nabla u(x))^T (I + \delta \nabla u(x)) = I + 2\delta \, \mathfrak{E}u(x) + \delta^2 \, \nabla u(x)^T \nabla u(x)$$

where

$$\mathfrak{E}u := \tfrac{1}{2}(\nabla u + (\nabla u)^T)$$

stands for the *symmetrized gradient*, i.e. the symmetric part of the gradient. Therefore, for $\delta \ll 1$ under a suitable differentiability assumption

$$V^{(\delta)}(x, Y) = \frac{1}{\delta^2} W(x, I + \delta Y) \approx \frac{1}{\delta^2} W(x, I + \delta Y_{\mathrm{sym}}) \approx \frac{1}{2} \partial_Y^2 W(x, I)[Y_{\mathrm{sym}}, Y_{\mathrm{sym}}].$$

We denote by $Y_{\mathrm{sym}} := \tfrac{1}{2}(Y + Y^T)$ the symmetrical part of Y and by $\mathbb{R}^{n \times n}_{\mathrm{sym}}$ the space of all symmetric matrices of dimension n. The forth-order tensor $\mathrm{A}(x) := \partial_Y^2 W(x, I)$ is called the *elasticity tensor* (at x). The corresponding energy given by the integral functional

$$\mathcal{E}^{(0)}(u) = \begin{cases} \tfrac{1}{2} \int_\Omega \mathrm{A}(x)[\mathfrak{E}u(x), \mathfrak{E}u(x)] \, dx, & u \in W^{1,2}(\Omega; \mathbb{R}^n), \\ \infty, & \text{else,} \end{cases}$$

is even quadratic. It was shown in [DNP:02] that this indeed is a good approximation of
the original energy

$$\mathcal{E}^{(\delta)}(u) := \begin{cases} \frac{1}{\delta^2} \int_\Omega W(x, I + \delta \nabla u(x)) \, dx, & u \in W^{1,2}(\Omega; \mathbb{R}^n), \\ \infty, & \text{else}, \end{cases}$$

since

$$\Gamma(L^2)\text{-}\lim_{\delta \to 0} \mathcal{E}^{(\delta)} = \mathcal{E}^{(0)}.$$

1.3   Geometric linearization in the multiwell case

In the previous section the stored-energy function had a simple one-well structure. However, also the case with multiple wells is physically relevant since, e.g., shape-memory
alloys in the martensite phase have such structure. For a survey of these materials, see
[Bh:06]. We consider a similar model as above with small displacements and additionally
with small distances between the wells, both of magnitude δ. For the complete discussion, including the justification of the choice of scales, we refer to [Sch:08] and references
therein.

Let for any $\delta > 0$

$$\tilde{\Sigma}_\delta := \bigcup_{S \in \Sigma} SO(n)(I + \delta S)$$

be the zero set of the stored-energy function $W^{(\delta)}$ where $\Sigma \subset \mathbb{R}^{n \times n}$ is a given finite subset
of positive matrices. (Our results, however, still hold if we merely assume that we have a
compact subset of symmetric matrices.) More precisely, for every $x \in \Omega$

$$W^{(\delta)}(x, X) = 0 \iff X \in \tilde{\Sigma}_\delta.$$

Obviously, for $\delta \ll 1$ all favourable states lie close to the reference configuration. Again,
in the most of the physically relevant cases, $W^{(\delta)}$ should grow quadratically with the
distance to the zero set. Here we allow for a more general p-growth for any $p > 1$, i.e.,
for some $C > 0$

$$W^{(\delta)}(x, X) \geq C \operatorname{dist}^p(X, \tilde{\Sigma}_\delta)$$

for all $x \in \Omega$ and $X \in \mathbb{R}^{n \times n}$. Regarding the displacements, we still rescale

$$V^{(\delta)} : \Omega \times \mathbb{R}^{n \times n}_{\text{sym}} \to \mathbb{R}, \quad V^{(\delta)}(x, Y) := \frac{1}{\delta^p} W^{(\delta)}(x, I + \delta Y).$$

However, contrary to the one-well case, we do not have an obvious candidate for the limit.
In order to make the transition from $\delta \ll 1$ to $\delta \to 0$, we must suppose that

- there exists a constant $C > 0$ such that

$$V^{(\delta)}(x, X) \leq C(|X|^p + 1)$$

  for all $x \in \Omega$ and $X \in \mathbb{R}^{n \times n}_{\text{sym}}$,

- the family $\{V^{(\delta)}\}_{\delta > 0}$ converges locally uniformly to some

$$V : \Omega \times \mathbb{R}^{n \times n}_{\text{sym}} \to \mathbb{R}$$

  as $\delta \to 0$.

It was shown in Theorem 2.1 of [Sch:08] for $p = 2$ that under the assumptions and denotations above, the family of functionals $\mathcal{E}^{(\delta)} : L^2(\Omega; \mathbb{R}^n) \to \mathbb{R}$, $\delta > 0$, given by

$$\mathcal{E}^{(\delta)}(u) := \begin{cases} \frac{1}{\delta^2} \int_\Omega W^{(\delta)}(x, I + \delta \nabla u(x)) \ dx, & u \in W^{1,2}(\Omega; \mathbb{R}^n), \\ \infty, & \text{else,} \end{cases}$$

Γ-converges in $L^2(\Omega; \mathbb{R}^n)$ as $\delta \to 0$ to

$$\mathcal{E}^{(\mathrm{rel})}(u) = \begin{cases} \int_\Omega V^{\mathrm{qcls}}(x, \mathfrak{E}u(x)) \ dx, & u \in W^{1,2}(\Omega; \mathbb{R}^n), \\ \infty, & \text{else.} \end{cases}$$

V^{qcls} denotes the symmetric-quasiconvex envelope of V (see Appendix A). The new problem is, though, still nonlinear; therefore we may speak just of geometric linearization.

The same result, with the necessary modifications, holds actually for every $p > 1$, as we will see in Section 3.3.

1.4   Homogenization

A material that has the same elastic properties in all of its points is said to be homogeneous. In this case the stored-energy function clearly does not depend on the location. By mixing two or more homogeneous materials, we may get a composite that better suits the intended usage. If the mixture is very fine and regular, it should behave on the macroscopic level approximately as a homogeneous material. Therefore, we are looking for a good approximation of the actual energy by a simpler homogeneous one.

We suppose that the material possesses a periodic cubic structure, and let $\varepsilon \ll 1$ denote the length of the pattern cell. Since we want to send $\varepsilon \to 0$, we give the stored-energy function as a $\mathbb{I}^n$-periodic function, $\mathbb{I} := (0, 1)$ denoting the unit interval, and appropriately rescale the actual position. Thus

$$\mathcal{E}_\varepsilon(y) = \int_\Omega W\left(\frac{x}{\varepsilon}, \nabla y(x)\right) \ dx$$

for an admissible deformation y. We assume $W : \mathbb{R}^n \times \mathbb{R}^{m \times n} \to \mathbb{R}$ to be Carathéodory (and $\mathbb{I}^n$-periodic) and for some $p > 1$ to fulfil the standard p-growth condition, i.e., there exist $\alpha, \beta > 0$ such that

$$\alpha |X|^p - \beta \le W(x, X) \le \beta(|X|^p + 1)$$

for a.e. $x \in \mathbb{R}^n$ and every $X \in \mathbb{R}^{m \times n}$. (In our case, $m = n$. The following result holds, though, also for $m \ne n$.) Under these conditions the family of functionals $\mathcal{E}_\varepsilon$, $\varepsilon > 0$, given by

$$\mathcal{E}_\varepsilon(y) := \begin{cases} \int_\Omega W(\frac{x}{\varepsilon}, \nabla y(x)) \ dx, & y \in W^{1,p}(\Omega; \mathbb{R}^m), \\ \infty, & \text{else,} \end{cases}$$

$\Gamma(L^p)$- converges to

$$\mathcal{E}_{\mathrm{hom}}(y) = \begin{cases} \int_\Omega W_{\mathrm{hom}}(\nabla y(x)) \ dx, & y \in W^{1,p}(\Omega; \mathbb{R}^m), \\ \infty, & \text{else.} \end{cases}$$

The homogenized stored-energy function is given by

$$W_{\text{hom}}(X) = \lim_{t \to \infty} \inf \left\{ \frac{1}{t^n} \int_{(0,t)^n} W(x, X + \nabla\varphi(x)) \, dx : \varphi \in W_0^{1,p}((0,t)^n; \mathbb{R}^m) \right\}.$$

The alternative expression is the multicell formula

$$W_{\text{hom}}(X) = \inf_{k \in \mathbb{N}} \inf \left\{ \frac{1}{k^n} \int_{k\mathbb{I}^n} W(x, X + \nabla\varphi(x)) \, dx : \varphi \in W_0^{1,p}(k\mathbb{I}^n; \mathbb{R}^m) \right\}.$$

The basic references for this result are [Br:85] and [Mü:87], see also the book [BD:98]. If W is convex in the second variable, the formula above reduces to the one-cell variant

$$W_{\text{hom}}(X) = \inf \left\{ \int_{\mathbb{I}^n} W(x, X + \nabla\varphi(x)) \, dx : \varphi \in W_0^{1,p}(\mathbb{I}^n; \mathbb{R}^m) \right\}.$$

This special case was solved already in [Ma:78].

Let us at this moment point out that for a location-independent function $W : \mathbb{R}^n \to \mathbb{R}$ homogenization reduces to quasiconvexification as

$$W_{\text{hom}}(X) = \inf_{k \in \mathbb{N}} \inf_{W_0^{1,p}(k\mathbb{I}^n; \mathbb{R}^m)} \frac{1}{k^n} \int_{k\mathbb{I}^n} W(X + \nabla\varphi(x)) \, dx = \inf_{k \in \mathbb{N}} W^{\text{qc}}(X) = W^{\text{qc}}(X).$$

1.5 Commutability of geometric linearization and homogenization

The last three sections imply the following questions: What would be a good effective theory when we have a fine mixture as well as small displacements? If we first homogenize, may we additionally linearize the new problem and thus get the desired result? What happens if we change the order of the two processes or do them simultaneously? These questions were resolved for the one-well case in [MN:11], where the authors prove that the two Γ-limits are indeed interchangeable. For a more general case including the two-scale convergence and simultaneous limits, see Chapter 5 in [Ne:10].

It will follow from our more general results that the two processes commute also in the multiwell case. Let us state all the assumptions and the result in detail.

Theorem (1.1). *Let $\Omega \subset \mathbb{R}^n$ be a bounded Lipschitz domain, $p > 1$ and $\Sigma \subset \mathbb{R}_{\text{sym}}^{n \times n}$ a finite set of positive matrices. Let us for every $\delta > 0$ have a Carathéodory function*

$$W^{(\delta)} : \mathbb{R}^n \times \mathbb{R}^{n \times n} \to \mathbb{R}$$

which

- *is $\mathbb{I}^n$-periodic in the first variable,*
- *is frame-indifferent, i.e., for a.e. $x \in \mathbb{R}^n$ and every $X \in \mathbb{R}^{n \times n}$*

$$W^{(\delta)}(x, RX) = W^{(\delta)}(x, X) \quad \text{for all } R \in \text{SO}(n),$$

- *has a multiwell structure at the scale δ, i.e., there exist $\alpha, \beta > 0$ such that for a.e. $x \in \mathbb{R}^n$*

$$\alpha \operatorname{dist}^p(X, \tilde{\Sigma}_\delta) \leq W^{(\delta)}(x, X) \leq \beta \operatorname{dist}^p(X, \tilde{\Sigma}_\delta) + \beta \delta^p$$

where

$$\tilde{\Sigma}_\delta := \bigcup_{S \in \Sigma} \mathrm{SO}(n)(I + \delta S).$$

Furthermore, we suppose that the functions

$$V^{(\delta)} : \mathbb{R}^n \times \mathbb{R}^{n \times n}_{\mathrm{sym}} \to \mathbb{R}, \quad V^{(\delta)}(x, Y) := \frac{1}{\delta^p} W^{(\delta)}(x, I + \delta Y),$$

converge locally uniformly to some Carathéodory function $V : \mathbb{R}^n \times \mathbb{R}^{n \times n}_{\mathrm{sym}} \to \mathbb{R}$. Let us for all $\delta, \varepsilon > 0$ define for $u \in W^{1,p}(\Omega; \mathbb{R}^n)$

$$
\begin{aligned}
\mathcal{E}_\varepsilon^{(\delta)}(u) &:= \frac{1}{\delta^p} \int_\Omega W^{(\delta)}\left(\frac{x}{\varepsilon}, I + \delta \nabla u(x)\right) \, dx, \\
\mathcal{E}_\varepsilon^{(\mathrm{lin})}(u) &:= \int_\Omega V^{\mathrm{qcls}}\left(\frac{x}{\varepsilon}, \mathfrak{E}u(x)\right) \, dx, \\
\mathcal{E}_{\mathrm{hom}}^{(\delta)}(u) &:= \frac{1}{\delta^p} \int_\Omega W_{\mathrm{hom}}^{(\delta)}\left(I + \delta \nabla u(x)\right) \, dx, \\
\mathcal{E}_{\mathrm{hom}}^{(\mathrm{lin})}(u) &:= \int_\Omega V_{\mathrm{hom}}\left(\mathfrak{E}u(x)\right) \, dx,
\end{aligned}
$$

and extend the definitions to $L^p(\Omega; \mathbb{R}^n)$ by ∞. V^{qcls} stands for the symmetric-quasiconvex envelope of V and

$$V_{\mathrm{hom}}(Y) := \inf_{k \in \mathbb{N}} \inf \left\{ \frac{1}{k^n} \int_{k\mathbb{I}^n} V(x, Y + \mathfrak{E}\varphi(x)) \, dx : \varphi \in W_0^{1,p}(k\mathbb{I}^n; \mathbb{R}^n) \right\}.$$

Then the following diagram commutes

$$
\begin{array}{ccc}
\mathcal{E}_\varepsilon^{(\delta)} & \xrightarrow[\Gamma(L^p)]{\delta \to 0} & \mathcal{E}_\varepsilon^{(\mathrm{lin})} \\
\Big\downarrow{\scriptstyle\varepsilon \to 0}\ \Gamma(L^p) & \Gamma(L^p) & \Gamma(L^p)\ \Big\downarrow{\scriptstyle\varepsilon \to 0} \\
\mathcal{E}_{\mathrm{hom}}^{(\delta)} & \xrightarrow[\delta \to 0]{\Gamma(L^p)} & \mathcal{E}_{\mathrm{hom}}^{(\mathrm{lin})}
\end{array}
$$

All the limits are $\Gamma(L^p)$-limits.

As already mentioned, this result will follow from a more general theory on the commutability of Γ-limits in the next chapter.

Chapter 2

The $p > 1$ case

In this central chapter of Part I, we will construct a general framework to deal with problems of similar kind as in Theorem (1.1). It will be called Γ-closure as an allusion to the homogenization closure, which was introduced and proved by Braides in [Br:86, Br:92] (see also Chapter 17 in [BD:98]). Compared to his works, we will start in a more general manner by taking any Γ-converging process instead of explicitly considering homogenization. Throughout the chapter we will assume $1 < p < \infty$.

Results from this chapter with their applications to elasticity theory, which will be presented in Chapter 3, are contained in our joint article with Schmidt [JS:14].

We will start by describing the aforementioned concept of Γ-closure. We will introduce an equivalence condition for families of functions that, sloppily said, corresponds to the local uniform convergence. Under the standard growth conditions this equivalence notion will turn out to be sufficient for Γ-closure results to hold. In Theorems (2.2) and (2.4) we will give and prove a version for a single domain as well as for variable ones, the latter meaning that we consider Γ-convergence for all subdomains of the given domain. A very important tool will be Lemma (C.4) that ensures the existence of modifications of bounded sequences in $W^{1,p}$ with p-equiintegrable gradients.

Since the motivating problem does not fall within the scope of functionals with a standard growth, we will in Section 2.3 relax the lower-bound inequality to some sort of Gårding's inequality. Accordingly, we will call such functionals to be of Gårding type. Nevertheless, the previous results will remain valid. In Section 2.4 we will end general considerations to Γ-closure by showing the complementary result with given boundary values. We will also recognize a perturbation and a relaxation result for Γ-convergence as immediate consequences of the derived theory.

In Section 2.6 we come to the question of commutability of Γ-limits. By assuming a stronger equivalence of families, we will be able to prove the order of taking two consecutive Γ-limits to be interchangeable. We will also address the related question regarding simultaneous limits.

Lastly, we will specialize to homogenization and extend the homogenization closure results of Braides. In accordance to his terminology, we will introduce homogenizable functions. However, in our setting the lower bound will have to be only of Gårding type, and we will not suppose any type of convexity. In addition, we will obtain a formula for the homogenized function.

2.1   Concept of Γ-closure

Let Ω be an open subset of $\mathbb{R}^n$ and $f_\varepsilon^{(j)} : \Omega \times \mathbb{R}^{m \times n} \to \mathbb{R}$ a (doubly indexed) family of Borel functions that are uniformly bounded from below. For $U \subset \Omega$ bounded and open, we define the corresponding integral functional

$$\mathcal{F}_\varepsilon^{(j)}(\_, U) : L^p(\Omega; \mathbb{R}^m) \to \mathbb{R} \cup \{\infty\}$$

by

$$\mathcal{F}_\varepsilon^{(j)}(u, U) := \left\{ \begin{array}{ll} \int_U f_\varepsilon^{(j)}(x, \nabla u(x)) \; dx, & u \in W^{1,p}(\Omega; \mathbb{R}^m), \\ \infty, & \text{else.} \end{array} \right.$$

If Ω itself is bounded, we simply write $\mathcal{F}_\varepsilon^{(j)}(\_, \Omega) =: \mathcal{F}_\varepsilon^{(j)}$. In view of our applications and also for notational clarity, we choose the indices $j \in \mathbb{N} \cup \{\infty\}$ and ε as the elements of a positive null sequence or 0.

Our main aim is to provide a rather general set of conditions that are easy to check in applications and that allow for a Γ-closure result of the following type: If

- the functionals $\mathcal{F}_\varepsilon^{(j)}$ Γ-converge as $\varepsilon \to 0$ for each finite j,

- the densities $f_\varepsilon^{(j)}$ are close to $f_\varepsilon^{(\infty)}$ for $\varepsilon > 0$ and large j in a suitable sense,

then

- also the Γ-limit for $j = \infty$ exists and

- is given as a limit as $j \to \infty$ of the Γ-limits for finite j.

2.2   Γ-closure under standard growth assumptions

As the first step, in this section we consider densities of standard p-growth. More precisely, we state the following

Definition (2.1).

(a) We say that the families $\{\{f_\varepsilon^{(j)}\}_{\varepsilon > 0}\}_{j \in \mathbb{N}}$ and $\{f_\varepsilon^{(\infty)}\}_{\varepsilon > 0}$ are *equivalent on an open subset* $U \subset \Omega$ if

$$\lim_{j \to \infty} \limsup_{\varepsilon \to 0} \int_U \sup_{|X| \le R} |f_\varepsilon^{(j)}(x, X) - f_\varepsilon^{(\infty)}(x, X)| \; dx = 0$$

for every $R \ge 0$.

(b) A family $f_\varepsilon^{(j)} : \Omega \times \mathbb{R}^{m \times n} \to \mathbb{R}$ of functions is said to *uniformly fulfil a standard p-growth condition* if there are $\alpha, \beta > 0$ independent of j and ε such that

$$\alpha |X|^p - \beta \le f_\varepsilon^{(j)}(x, X) \le \beta(|X|^p + 1)$$

for a.e. $x \in \Omega$ and all $X \in \mathbb{R}^{m \times n}$.

For a single function we will also simply say that it has a (standard) p-growth. Rather than of 1- and 2-growth, we will speak of linear and quadratic growth.

Theorem (2.2). *Let $\Omega \subset \mathbb{R}^n$ be open and bounded. Suppose that the family of Borel functions*

$$f_\varepsilon^{(j)} : \Omega \times \mathbb{R}^{m \times n} \to \mathbb{R}, \quad j \in \mathbb{N} \cup \{\infty\}, \; \varepsilon > 0,$$

uniformly fulfils a standard p-growth condition, and let $\mathcal{F}_\varepsilon^{(j)} : L^p(\Omega; \mathbb{R}^m) \to \mathbb{R} \cup \{\infty\}$ be the corresponding integral functionals. Assume that

- *for each $j < \infty$ the Γ-limit $\Gamma(L^p)\text{-}\lim_{\varepsilon \to 0} \mathcal{F}_\varepsilon^{(j)} =: \mathcal{F}_0^{(j)}$ exists,*

- *the families $\big\{ \{ f_\varepsilon^{(j)} \}_{\varepsilon > 0} \big\}_{j \in \mathbb{N}}$ and $\{ f_\varepsilon^{(\infty)} \}_{\varepsilon > 0}$ are equivalent on Ω.*

Then also $\Gamma(L^p)\text{-}\lim_{\varepsilon \to 0} \mathcal{F}_\varepsilon^{(\infty)} =: \mathcal{F}_0^{(\infty)}$ exists and is the pointwise as well as the Γ-limit of $\mathcal{F}_0^{(j)}$ as $j \to \infty$:

$$\mathcal{F}_0^{(\infty)} = \lim_{j \to \infty} \mathcal{F}_0^{(j)} = \Gamma(L^p)\text{-}\lim_{j \to \infty} \mathcal{F}_0^{(j)}.$$

Remark (2.3).

(a) Schematically, this theorem can be summarized by the diagram below. The assumptions are given by solid lines, where in particular the equivalence of densities is indicated by a double line. The consequences of the theorem are given by dashed arrows.

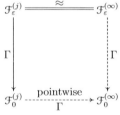

(b) Note that by Theorem (B.4) also the functionals $\mathcal{F}_0^{(j)}$, $j \in \mathbb{N} \cup \{\infty\}$, are integral functionals with Carathéodory densities $f_0^{(j)}$ of standard p-growth. In fact, being densities of Γ-limits, the $f_0^{(j)}$ are quasiconvex in the second argument.

(c) Our definition of equivalence of the sequences $\{ \{ f_\varepsilon^{(j)} \}_{\varepsilon > 0} \}_{j \in \mathbb{N}}$ and $\{ f_\varepsilon^{(\infty)} \}_{\varepsilon > 0}$ is motivated by the notion of equivalence in [Br:86], which will appear in Theorem (2.31).

The previous result can be extended to functionals on variable domains in a straightforward manner. For $\Omega \subset \mathbb{R}^n$ open (not necessarily bounded), denote by $\mathcal{A}(\Omega)$ the set of all bounded open subsets of Ω with Lipschitz boundary.

Theorem (2.4). *Let $\Omega \subset \mathbb{R}^n$ be open. Suppose that a family of Borel functions*

$$f_\varepsilon^{(j)} : \Omega \times \mathbb{R}^{m \times n} \to \mathbb{R}, \quad j \in \mathbb{N} \cup \{\infty\}, \; \varepsilon > 0$$

uniformly fulfils a standard p-growth condition. Assume for every $U \in \mathcal{A}(\Omega)$ that

- for each $j < \infty$ the Γ-limit $\Gamma(L^p)$-$\lim_{\varepsilon \to 0} \mathcal{F}_\varepsilon^{(j)}(\,\_\,, U)$ exists, $\mathcal{F}_\varepsilon^{(j)}$ being the corresponding integral functionals,

- the families $\{\{f_\varepsilon^{(j)}\}_{\varepsilon > 0}\}_{j \in \mathbb{N}}$ and $\{f_\varepsilon^{(\infty)}\}_{\varepsilon > 0}$ are equivalent on U.

Then for each $j \in \mathbb{N} \cup \{\infty\}$ there exists a Carathéodory function $f_0^{(j)} : \Omega \times \mathbb{R}^{m \times n} \to \mathbb{R}$, uniquely determined a.e. on Ω, such that for the corresponding integral functional $\mathcal{F}_0^{(j)}$ it holds

$$\Gamma(L^p)\text{-}\lim_{\varepsilon \to 0} \mathcal{F}_\varepsilon^{(j)}(\,\_\,, U) = \mathcal{F}_0^{(j)}(\,\_\,, U)$$

for all $U \in \mathcal{A}(\Omega)$. Moreover, for every $u \in L^p(\Omega; \mathbb{R}^m)$

$$\mathcal{F}_0^{(\infty)}(u, U) = \lim_{j \to \infty} \mathcal{F}_0^{(j)}(u, U) = \Gamma(L^p)\text{-}\lim_{j \to \infty} \mathcal{F}_0^{(j)}(u, U),$$

and the limiting densities $f_0^{(j)}$ satisfy for all $X \in \mathbb{R}^{m \times n}$

$$f_0^{(j)}(\,\_\,, X) \overset{*}{\rightharpoonup} f_0^{(\infty)}(\,\_\,, X) \quad \text{in } L^\infty(\Omega).$$

Remark (2.5). It is worth pointing out that for any family $\{\{f_\varepsilon^{(j)}\}_{\varepsilon > 0}\}_{j \in \mathbb{N}}$ there is always a subsequence $\{\varepsilon_k\}_{k \in \mathbb{N}}$ of ε such that $\Gamma(L^p)$-$\lim_{\varepsilon \to 0} \mathcal{F}_\varepsilon^{(j)}(\,\_\,, U)$ exists for all $j \in \mathbb{N}$ and $U \in \mathcal{A}(\Omega)$. This follows from Theorem (B.4) in combination with a standard diagonal sequence argument.

Proof (of Theorem (2.2)). The proof is divided into three steps. In the first two we assume that

- $\Gamma(L^p)$-$\lim_{\varepsilon \to 0} \mathcal{F}_\varepsilon^{(\infty)} =: \mathcal{F}_0^{(\infty)}$ exists,

- $\mathcal{F}_0^{(\infty)}(u) < \infty \iff u \in W^{1,p}(U; \mathbb{R}^m)$.

This will be justified in Step 3.

Step 1: Upper bound. For $u \in L^p(\Omega; \mathbb{R}^m)$ we claim that

$$(2.6) \qquad\qquad \limsup_{j \to \infty} \mathcal{F}_0^{(j)}(u) \leq \mathcal{F}_0^{(\infty)}(u).$$

This is obvious if $u \in L^p(\Omega; \mathbb{R}^m) \setminus W^{1,p}(\Omega; \mathbb{R}^m)$. If $u \in W^{1,p}(\Omega; \mathbb{R}^m)$, we choose a recovery sequence $\{u_\varepsilon\}_\varepsilon$ for u with $\mathcal{F}_\varepsilon^{(\infty)}(u_\varepsilon) \to \mathcal{F}_0^{(\infty)}(u)$. From the lower bound it follows that $\{u_\varepsilon\}_\varepsilon$ contains a subsequence that is bounded in $W^{1,p}(\Omega; \mathbb{R}^m)$. Hence, by Lemma (C.4) there exists a further subsequence $\{u_{\varepsilon_k}\}_{k \in \mathbb{N}}$ and functions $v_k \in W^{1,p}(\Omega; \mathbb{R}^m)$ such that

- $\{|\nabla v_k|^p\}_{k \in \mathbb{N}}$ is equiintegrable on Ω,

- $|A_k| \to 0$ for $A_k := \{x \in \Omega : \nabla u_{\varepsilon_k}(x) \neq \nabla v_k(x)\}$ and

- $v_k \rightharpoonup u$ in $W^{1,p}(\Omega; \mathbb{R}^m)$.

By setting $E_k^M := \{x \in \Omega : |\nabla v_k(x)| \geq M\}$, we obtain that

$$
\begin{aligned}
\mathcal{F}_{\varepsilon_k}^{(\infty)}(u_{\varepsilon_k}) &\geq \int_{\Omega \setminus (A_k \cup E_k^M)} f_{\varepsilon_k}^{(\infty)}(x, \nabla v_k(x)) \, dx - \beta|A_k \cup E_k^M| \\
&\geq \int_{\Omega \setminus E_k^M} f_{\varepsilon_k}^{(\infty)}(x, \nabla v_k(x)) \, dx - |A_k \setminus E_k^M|\beta(M^p + 1) - \beta|A_k \cup E_k^M| \\
&\geq \int_{\Omega \setminus E_k^M} f_{\varepsilon_k}^{(j)}(x, \nabla v_k(x)) \, dx - \int_\Omega \sup_{|X| \leq M} |f_{\varepsilon_k}^{(j)}(x, X) - f_{\varepsilon_k}^{(\infty)}(x, X)| \, dx - \\
&\quad - |A_k|\beta(M^p + 2) - \beta|E_k^M| \\
&\geq \mathcal{F}_{\varepsilon_k}^{(j)}(v_k) - \int_{E_k^M} \beta(|\nabla v_k(x)|^p + 1) \, dx - \\
&\quad - \int_\Omega \sup_{|X| \leq M} |f_{\varepsilon_k}^{(j)}(x, X) - f_{\varepsilon_k}^{(\infty)}(x, X)| \, dx - |A_k|\beta(M^p + 2) - \beta|E_k^M|.
\end{aligned}
$$

As $\{|\nabla v_k|^p\}_{k \in \mathbb{N}}$ is equiintegrable, we may for given $\eta > 0$ find M so large that for all k it holds $\beta \int_{E_k^M}(|\nabla v_k(x)|^p + 2) \, dx \leq \eta$. Letting first $k \to \infty$, we find that

$$
\begin{aligned}
\mathcal{F}_0^{(\infty)}(u) &= \lim_{k \to \infty} \mathcal{F}_{\varepsilon_k}^{(\infty)}(u_{\varepsilon_k}) \\
&\geq \limsup_{k \to \infty} \mathcal{F}_{\varepsilon_k}^{(j)}(v_k) - \eta - \limsup_{k \to \infty} \int_\Omega \sup_{|X| \leq M} |f_{\varepsilon_k}^{(j)}(x, X) - f_{\varepsilon_k}^{(\infty)}(x, X)| \, dx.
\end{aligned}
$$

Now let $j \to \infty$. Due to our equivalence assumption, and since $\eta > 0$ was arbitrary, we indeed arrive at

$$
\mathcal{F}_0^{(\infty)}(u) \geq \limsup_{j \to \infty} \limsup_{k \to \infty} \mathcal{F}_{\varepsilon_k}^{(j)}(v_k) \geq \limsup_{j \to \infty} \mathcal{F}_0^{(j)}(u)
$$

by the lim inf-inequality for $\Gamma(L^p)\text{-}\lim_{k \to \infty} \mathcal{F}_{\varepsilon_k}^{(j)} = \mathcal{F}_0^{(j)}$.

Step 2: Lower bound. We claim that

$$
\tag{2.7} \liminf_{j \to \infty} \mathcal{F}_0^{(j)}(u_j) \geq \mathcal{F}_0^{(\infty)}(u)
$$

whenever $u_j \to u$ in $L^p(\Omega; \mathbb{R}^m)$.

To prove this, we pass to a subsequence $\{j_k\}_{k \in \mathbb{N}}$ such that

$$
\lim_{k \to \infty} \mathcal{F}_0^{(j_k)}(u_{j_k}) = \liminf_{j \to \infty} \mathcal{F}_0^{(j)}(u_j).
$$

We may without loss of generality assume that $\{\mathcal{F}_0^{(j_k)}(u_{j_k})\}_{k \in \mathbb{N}}$ is bounded and that j_k is chosen so large that

$$
\limsup_{\varepsilon \to 0} \int_\Omega \sup_{|X| \leq k} |f_\varepsilon^{(j_k)}(x, X) - f_\varepsilon^{(\infty)}(x, X)| \, dx < \frac{1}{k}.
$$

Then we choose ε_{j_k} (with $\varepsilon_{j_k} \searrow 0$) so small that there is a $w_{j_k} \in L^p(\Omega; \mathbb{R}^m)$ with

$$
\|w_{j_k} - u_{j_k}\|_{L^p} \leq \frac{1}{j_k} \quad \text{and} \quad \mathcal{F}_0^{(j_k)}(u_{j_k}) + \frac{1}{j_k} \geq \mathcal{F}_{\varepsilon_{j_k}}^{(j_k)}(w_{j_k})
$$

as well as

$$\int_\Omega \sup_{|X| \le k} |f^{(j_k)}_{\varepsilon_{j_k}}(x, X) - f^{(\infty)}_{\varepsilon_{j_k}}(x, X)| \, dx$$

$$\le \tfrac{1}{k} + \limsup_{\varepsilon \to 0} \int_\Omega \sup_{|X| \le k} |f^{(\infty)}_\varepsilon(x, X) - f^{(\infty)}_\varepsilon(x, X)| \, dx$$

$$\le \tfrac{2}{k}.$$

By the lower bound and Lemma (C.4), there is a further subsequence $\{j_{k_i}\}_{i \in \mathbb{N}}$ and a sequence $\{v_i\}_{i \in \mathbb{N}} \in W^{1,p}(\Omega; \mathbb{R}^m)$ such that

- $\{|\nabla v_i|^p\}_{i \in \mathbb{N}}$ is equiintegrable,

- $|A_i| \to 0$ for $A_i := \{x \in \Omega : \nabla w_{j_{k_i}}(x) \ne \nabla v_i(x)\}$,

- $v_i \rightharpoonup u$ in $W^{1,p}(\Omega; \mathbb{R}^m)$.

Then for $E_i^M := \{x \in \Omega : |\nabla v_i(x)| \ge M\}$, we have

$$
\begin{aligned}
\mathcal{F}^{(j_{k_i})}_{\varepsilon_{j_{k_i}}}(w_{j_{k_i}}) &= \int_\Omega f^{(j_{k_i})}_{\varepsilon_{j_{k_i}}}(x, \nabla w_{j_{k_i}}) \\
&\ge \int_{\Omega \setminus (A_i \cup E_i^M)} f^{(j_{k_i})}_{\varepsilon_{j_{k_i}}}(x, \nabla v_i(x)) \, dx - \beta |A_i \cup E_i^M| \\
&\ge \int_{\Omega \setminus E_i^M} f^{(j_{k_i})}_{\varepsilon_{j_{k_i}}}(x, \nabla v_i(x)) \, dx - |A_i \setminus E_i^M| \beta(M^p + 1) - \beta |A_i \cup E_i^M| \\
&\ge \int_{\Omega \setminus E_i^M} f^{(\infty)}_{\varepsilon_{j_{k_i}}}(x, \nabla v_i(x)) \, dx - \int_\Omega \sup_{|X| \le M} |f^{(j_{k_i})}_{\varepsilon_{j_{k_i}}}(x, X) - f^{(\infty)}_{\varepsilon_{j_{k_i}}}(x, X)| \, dx - \\
&\quad - |A_i| \beta(M^p + 2) - \beta |E_i^M| \\
&\ge \mathcal{F}^{(\infty)}_{\varepsilon_{j_{k_i}}}(v_i) - \int_{E_i^M} \beta(|\nabla v_i|^p + 1) \, dx - \tfrac{2}{k_i} - |A_i| \beta(M^p + 2) - \beta |E_i^M|
\end{aligned}
$$

if $k_i \ge M$. For given $\eta > 0$, by choosing M large and then letting $i \to \infty$, we arrive at

$$\lim_{i \to \infty} \mathcal{F}^{(j_{k_i})}_0(u_{j_{k_i}}) \ge \limsup_{i \to \infty} \mathcal{F}^{(j_{k_i})}_{\varepsilon_{j_{k_i}}}(w_{j_{k_i}}) \ge \limsup_{i \to \infty} \mathcal{F}^{(\infty)}_{\varepsilon_{j_{k_i}}}(v_i) - \eta.$$

Since $v_i \to u$ in $L^p(\Omega; \mathbb{R}^m)$ and η was arbitrary,

$$\liminf_{j \to \infty} \mathcal{F}^{(j)}_0(u_j) = \lim_{i \to \infty} \mathcal{F}^{(j_{k_i})}_0(u_{j_{k_i}}) \ge \mathcal{F}^{(\infty)}_0(u),$$

as claimed.

Combining (2.6) and (2.7) with $u_j = u$ for all j, we arrive at

$$(2.8) \qquad\qquad \lim_{j \to \infty} \mathcal{F}^{(j)}_0(u) = \mathcal{F}^{(\infty)}_0(u)$$

for every $u \in L^p(\Omega; \mathbb{R}^m)$.

Step 3: Justification of our assumption.

If we do not assume a priori that $\mathcal{F}_\varepsilon^{(\infty)}$ Γ-converges to $\mathcal{F}_0^{(\infty)}$, by Theorem (B.4), for every subsequence $\{\varepsilon_k\}_{k\in\mathbb{N}}$ there exists a further subsequence $\{\varepsilon_{k_i}\}_{i\in\mathbb{N}}$ such that

$$\Gamma(L^p)\text{-}\lim_{i\to\infty} \mathcal{F}_{\varepsilon_{k_i}}^{(\infty)} =: \mathcal{F}_0^{(\infty)}$$

exists and takes finite values precisely on $W^{1,p}(\Omega;\mathbb{R}^m)$. Proceeding as above, we infer from (2.8) that $\mathcal{F}_0^{(\infty)}$ does not depend on the particular subsequence $\{\varepsilon_{k_i}\}_{i\in\mathbb{N}}$. Employing the Urysohn property for Γ-limits (see Theorem (B.3)), we thus find that indeed $\Gamma(L^p)\text{-}\lim_{\varepsilon\to 0} \mathcal{F}_\varepsilon^{(\infty)} = \mathcal{F}_0^{(\infty)}$. ∎

Proof (of Theorem (2.4)). Applying Theorem (2.2) to the functionals $\mathcal{F}_\varepsilon^{(j)}(\_,U)$ with $U \in \mathcal{A}(\Omega)$ fixed, we have for each $j \in \mathbb{N}\cup\{\infty\}$ a uniquely determined functional

$$\Gamma(L^p)\text{-}\lim_{\varepsilon\to 0} \mathcal{F}_\varepsilon^{(j)}(\_,U) =: \mathcal{G}_U^{(j)}.$$

On the other hand, by Theorem (B.4), for each j there exists a subsequence $\{\varepsilon_k\}_{k\in\mathbb{N}}$ such that

$$\mathcal{G}_U^{(j)} = \Gamma(L^p)\text{-}\lim_{k\to\infty} \mathcal{F}_{\varepsilon_k}^{(j)}(\_,U) = \mathcal{F}_0^{(j)}(\_,U) \quad \text{for all } U \in \mathcal{A}(\Omega)$$

for an integral functional $\mathcal{F}_0^{(j)} : L^p(\Omega;\mathbb{R}^m)\times\mathcal{A}(\Omega) \to \mathbb{R}\cup\{\infty\}$ with a Carathéodory density $f_0^{(j)}$. Hence $\mathcal{G}_U^{(j)} = \mathcal{F}_0^{(j)}(\_,U)$ for all $U \in \mathcal{A}(\Omega)$. By Theorem (2.2) we indeed have

$$\mathcal{F}_0^{(\infty)}(u,U) = \lim_{j\to\infty} \mathcal{F}_0^{(j)}(u,U) = \Gamma(L^p)\text{-}\lim_{j\to\infty} \mathcal{F}_0^{(j)}(u,U)$$

for all $u \in L^p(\Omega;\mathbb{R}^m)$ and $U \in \mathcal{A}(\Omega)$. Being continuous in the second argument, the densities $f_0^{(j)}$ are uniquely determined almost everywhere on Ω and everywhere on $\mathbb{R}^{m\times n}$ by

$$\int_U f_0^{(j)}(x,X)\,dx = \mathcal{F}_0^{(j)}(\ell_X,U) \quad \text{for every } U \in \mathcal{A}(\Omega),$$

where $\ell_X(x) := Xx$ for $X \in \mathbb{R}^{m\times n}$. The pointwise convergence yields for every $X \in \mathbb{R}^{m\times n}$ also

$$\lim_{j\to\infty}\int_U f_0^{(j)}(x,X)\,dx = \lim_{j\to\infty}\mathcal{F}_0^{(j)}(\ell_X,U) = \mathcal{F}_0^{(\infty)}(\ell_X,U) = \int_U f_0^{(\infty)}(x,X)\,dx$$

for all $U \in \mathcal{A}(\Omega)$. Since $\{f_0^{(j)}(\_,X)\}_{j\in\mathbb{N}}$ is uniformly bounded, it follows

$$f_0^{(j)}(\_,X) \overset{*}{\rightharpoonup} f_0^{(\infty)}(\_,X) \quad \text{in } L^\infty(\Omega). \quad \blacksquare$$

2.3 Γ-closure for Gårding type functionals

For many interesting applications (e.g., the ones to be discussed in Chapter 3), a standard p-growth assumption is too restrictive. In this section we generalize our Γ-closure theorem to integral functionals of "Gårding type". More precisely, while imposing p-growth assumptions from above as before, the integral densities will only assumed to be bounded from below by some constant. Yet the functionals are still supposed to satisfy a weak coercivity assumption, which we impose by requiring a Gårding type inequality to hold.

Definition (2.9). We say that a doubly indexed family of integral functionals

$$\mathcal{F}^{(j)}_{\varepsilon} : L^p(\Omega; \mathbb{R}^m) \to \mathbb{R} \cup \{\infty\}, \quad j \in \mathbb{N} \cup \{\infty\}, \ \varepsilon > 0,$$

corresponding to Borel functions $f^{(j)}_{\varepsilon} : \Omega \times \mathbb{R}^{m \times n} \to \mathbb{R}$, is *of uniform p-Gårding type on* $U \subset \Omega$ open if

- there is $\beta > 0$ independent of j and ε such that the $f^{(j)}_{\varepsilon}$ satisfy

$$-\beta \le f^{(j)}_{\varepsilon}(x, X) \le \beta(|X|^p + 1)$$

  for a.e. $x \in \Omega$ and all $X \in \mathbb{R}^{m \times n}$,

- there are $\alpha_U > 0$ and $\gamma_U \in \mathbb{R}$ such that

$$\mathcal{F}^{(j)}_{\varepsilon}(u) \ge \alpha_U \int_U |\nabla u(x)|^p \, dx - \gamma_U \int_U |u(x)|^p \, dx$$

  for all $u \in W^{1,p}(U; \mathbb{R}^m)$.

Clearly, we may have only one index. For a single integral functional, we will just say that it is of p-Gårding type on U.

Theorem (2.10). *Let $\Omega \subset \mathbb{R}^n$ be bounded and open. Suppose that the family of functionals $\mathcal{F}^{(j)}_{\varepsilon}$, $j \in \mathbb{N} \cup \{\infty\}$, $\varepsilon > 0$, with densities $f^{(j)}_{\varepsilon} : \Omega \times \mathbb{R}^{m \times n} \to \mathbb{R}$ is of uniform p-Gårding type on Ω. Assume that*

- *for each $j < \infty$ the Γ-limit $\Gamma(L^p)$-$\lim_{\varepsilon \to 0} \mathcal{F}^{(j)}_{\varepsilon} =: \mathcal{F}^{(j)}_0$ exists,*

- *the families $\left\{ \{f^{(j)}_{\varepsilon}\}_{\varepsilon > 0} \right\}_{j \in \mathbb{N}}$ and $\{f^{(\infty)}_{\varepsilon}\}_{\varepsilon > 0}$ are equivalent on Ω.*

Then also $\Gamma(L^p)$-$\lim_{\varepsilon \to 0} \mathcal{F}^{(\infty)}_{\varepsilon} =: \mathcal{F}^{(\infty)}_0$ exists and is the pointwise and the Γ-limit of $\mathcal{F}^{(j)}_0$ as $j \to \infty$:

$$\mathcal{F}^{(\infty)}_0 = \lim_{j \to \infty} \mathcal{F}^{(j)}_0 = \Gamma(L^p)\text{-}\lim_{j \to \infty} \mathcal{F}^{(j)}_0.$$

Remark (2.11). In fact, the assumption that the $f^{(j)}_{\varepsilon}$ are bounded from above can be dropped for $j < \infty$. In order to see this, it suffices to combine Lemma (C.4) with Proposition A.1 in [FJM:02] so as to obtain approximations with uniformly bounded gradients in the proof of Theorem (2.2). In general, one then only has that $\mathcal{F}^{(\infty)}_0 = \lim_{j \to \infty} \mathcal{F}^{(j)}_0$ on $W^{1,\infty}(\Omega; \mathbb{R}^m)$. By density, however, this is enough to obtain that still $\mathcal{F}^{(\infty)}_0 = \Gamma(L^p)$-$\lim_{j \to \infty} \mathcal{F}^{(j)}_0$. Although of interest in models of elasticity theory, we do not pursue this line of thought here as in our main application to homogenization theory in Section 2.7 the assumptions are only known to be satisfied under a standard p-growth assumption from above.

Again we also state a version of this result on variable domains. Note that here the constants in the Gårding inequality are allowed to explicitly depend upon the domain $U \subset \Omega$.

Theorem (2.12). *Let $\Omega \subset \mathbb{R}^n$ be open, and let us have a family of functionals $\mathcal{F}_\varepsilon^{(j)}$, $j \in \mathbb{N} \cup \{\infty\}$, $\varepsilon > 0$, with densities $f_\varepsilon^{(j)} : \Omega \times \mathbb{R}^{m \times n} \to \mathbb{R}$. We assume for every $U \in \mathcal{A}(\Omega)$ that*

- *the family $\mathcal{F}_\varepsilon^{(j)}$, $j \in \mathbb{N} \cup \{\infty\}$, $\varepsilon > 0$, is of uniform p-Gårding type on U,*

- *for each $j < \infty$ the Γ-limit $\Gamma(L^p)\text{-}\lim_{\varepsilon \to 0} \mathcal{F}_\varepsilon^{(j)}(\_, U)$ exists,*

- *the families $\left\{ \{ f_\varepsilon^{(j)} \}_{\varepsilon > 0} \right\}_{j \in \mathbb{N}}$ and $\{ f_\varepsilon^{(\infty)} \}_{\varepsilon > 0}$ are equivalent on U.*

Then for each $j \in \mathbb{N} \cup \{\infty\}$, there exists a Carathéodory function $f_0^{(j)} : \Omega \times \mathbb{R}^{m \times n} \to \mathbb{R}$, uniquely determined a.e. on Ω, such that for the corresponding integral functional $\mathcal{F}_0^{(j)}$ it holds

$$\Gamma(L^p)\text{-}\lim_{\varepsilon \to 0} \mathcal{F}_\varepsilon^{(j)}(\_, U) = \mathcal{F}_0^{(j)}(\_, U)$$

for all $U \in \mathcal{A}(\Omega)$. Moreover, for $u \in L^p(\Omega; \mathbb{R}^m)$

$$\mathcal{F}_0^{(\infty)}(u, U) = \lim_{j \to \infty} \mathcal{F}_0^{(j)}(u, U) = \Gamma(L^p)\text{-}\lim_{j \to \infty} \mathcal{F}_0^{(j)}(u, U),$$

and the limiting densities $f_0^{(j)}$ satisfy for all $X \in \mathbb{R}^{m \times n}$

$$f_0^{(j)}(\_, X) \overset{*}{\rightharpoonup} f_0^{(\infty)}(\_, X) \quad in \ L^\infty(\Omega).$$

Remark (2.13). The following proposition not only is the first step towards the proofs of Theorems (2.10) and (2.12). It also provides a criterion for the Γ-convergence of Gårding type functionals and might thus be used to verify the first assumption in Theorems (2.10) and (2.12) in particular situations. In particular, it shows that the assumptions of these theorems are always satisfied for suitable subsequences.

Proposition (2.14). *Let $\Omega \subset \mathbb{R}^n$ be open. Suppose that the family of functionals $\mathcal{F}_\varepsilon$, $\varepsilon > 0$, with densities $f_\varepsilon : \Omega \times \mathbb{R}^{m \times n} \to \mathbb{R}$ is of uniform p-Gårding type on every $U \in \mathcal{A}(\Omega)$. For arbitrary $\lambda_k \searrow 0$, define*

$$f_\varepsilon^{(k)}(x, X) := f_\varepsilon(x, X) + \lambda_k |X|^p,$$

and by $\mathcal{F}_\varepsilon^{(k)}$ denote the corresponding integral functionals. Assume that for every $k \in \mathbb{N}$ and all $U \in \mathcal{A}(\Omega)$ the Γ-limit

$$\Gamma(L^p)\text{-}\lim_{\varepsilon \to 0} \mathcal{F}_\varepsilon^{(k)}(\_, U) =: \mathcal{F}_0^{(k)}(\_, U)$$

exists. Then also

$$\Gamma(L^p)\text{-}\lim_{\varepsilon \to 0} \mathcal{F}_\varepsilon(\_, U) =: \mathcal{F}_0(\_, U)$$

exists and is given by

$$\mathcal{F}_0(\_, U) = \inf_{k \in \mathbb{N}} \mathcal{F}_0^{(k)}(\_, U) = \lim_{k \to \infty} \mathcal{F}_0^{(k)}(\_, U) = \Gamma(L^p)\text{-}\lim_{k \to \infty} \mathcal{F}_0^{(k)}(\_, U).$$

Moreover, $\mathcal{F}_0$ and $\mathcal{F}_0^{(k)}$, $j \in \mathbb{N}$, are given in terms of Carathéodory integral densities f_0 and $f_0^{(k)}$, respectively, and for a.e. $x \in \Omega$ and all $X \in \mathbb{R}^{m \times n}$ it holds

$$f_0(x, X) = \inf_{k \in \mathbb{N}} f_0^{(k)}(x, X) = \lim_{k \to \infty} f_0^{(k)}(x, X).$$

Proof. Since $\|\nabla u\|^p_{L^p(U)} \leq \alpha_U^{-1}(\mathcal{F}_\varepsilon(u,U) + \gamma_U \|u\|^p_{L^p(U)})$ for $u \in W^{1,p}(U;\mathbb{R}^m)$, we obtain

$$\left(1 - \frac{\lambda_k}{\alpha_U}\right)\mathcal{F}^{(k)}_\varepsilon(u,U) - \frac{\gamma_U \lambda_k}{\alpha_U}\|u\|^p_{L^p(U)} \leq \mathcal{F}_\varepsilon(u,U) \leq \mathcal{F}^{(k)}_\varepsilon(u,U)$$

for all $u \in L^p(U;\mathbb{R}^m)$. As the Γ-$\liminf$ and the Γ-$\limsup$ are stable under continuous perturbations, this implies that

$$(2.15) \qquad \left(1 - \frac{\lambda_k}{\alpha_U}\right)\mathcal{F}^{(k)}_0(u,U) - \frac{\gamma_U \lambda_k}{\alpha_U}\|u\|^p_{L^p(U)} \quad \leq \quad \Gamma(L^p)\text{-}\liminf_{\varepsilon \to 0} \mathcal{F}_\varepsilon(u,U),$$

$$(2.16) \qquad\qquad\qquad \Gamma(L^p)\text{-}\limsup_{\varepsilon \to 0} \mathcal{F}_\varepsilon(u,U) \quad \leq \quad \mathcal{F}^{(k)}_0(u,U),$$

and so, due to monotonicity in k,

$$(2.17) \qquad \mathcal{F}_0(u,U) := \Gamma(L^p)\text{-}\lim_{\varepsilon \to 0}\mathcal{F}_\varepsilon(u,U) = \inf_{k \in \mathbb{N}}\mathcal{F}^{(k)}_0(u,U) = \lim_{k \to \infty}\mathcal{F}^{(k)}_0(u,U).$$

Since $\mathcal{F}^{(k)}_0$ is decreasing in k and $\mathcal{F}_0(\_,U)$ is lower semicontinuous, also

$$\mathcal{F}_0(u,U) = \Gamma(L^p)\text{-}\lim_{k \to \infty}\mathcal{F}^{(k)}_0(u,U),$$

see Appendix B.

In order to prove the statement on the densities, we first note that by Theorem (B.4) there exist Carathéodory functions $f^{(k)}_0$ such that $\mathcal{F}^{(k)}_0(u,U) = \int_U f^{(k)}_0(x,\nabla u(x))\,dx$ for $u \in W^{1,p}(U;\mathbb{R}^m)$ and ∞ otherwise. By monotonicity in k, if $l \geq k$,

$$\int_U f^{(l)}_0(x,X)\,dx \leq \int_U f^{(k)}_0(x,X)\,dx$$

for all $X \in \mathbb{R}^{m \times n}$ and $U \in \mathcal{A}(\Omega)$. From continuity in the second argument, it follows for a.e. $x \in \Omega$

$$f^{(l)}_0(x,X) \leq f^{(k)}_0(x,X) \quad \text{for all } X \in \mathbb{R}^{m \times n}.$$

By the dominated convergence theorem it thus follows that

$$\mathcal{F}_0(u,U) = \lim_{k \to \infty}\mathcal{F}^{(k)}_0(u,U) = \int_U \lim_{k \to \infty} f^{(k)}_0(x,\nabla u(x))\,dx$$

for $u \in W^{1,p}(U;\mathbb{R}^m)$ and $\mathcal{F}_0(u,U) = +\infty$ otherwise. As $f^{(k)}_0(x,\_)$ is quasiconvex for almost every x, so is $f_0(x,\_)$, which shows that f_0 is Carathéodory. ∎

Remark (2.18). The first part of this proof shows that requiring the uniform p-Gårding assumption and the existence of the Γ-limits only on a single bounded and open domain Ω, one still has

$$\Gamma(L^p)\text{-}\lim_{\varepsilon \to 0}\mathcal{F}_\varepsilon = \inf_{j \in \mathbb{N}}\mathcal{F}^{(j)}_0 = \lim_{j \to \infty}\mathcal{F}^{(j)}_0.$$

We now prove Theorem (2.12) by using Proposition (2.14) and thus getting families with standard growth conditions. The proof of Theorem (2.10) will then be a straightforward adaptation of the first part of the following proof.

Proof (of Theorem (2.12)). Let for every $k \in \mathbb{N}$, $j \in \mathbb{N} \cup \{\infty\}$ and $\varepsilon > 0$

$$f_\varepsilon^{(k,j)}(x,X) := f_\varepsilon^{(j)}(x,X) + \frac{1}{k}|X|^p \quad \text{and} \quad \mathcal{F}_\varepsilon^{(k,j)}(u,U) := \mathcal{F}_\varepsilon^{(j)}(u,U) + \frac{1}{k}\int_U |\nabla u(x)|^p \, dx.$$

Clearly, for each fixed $k \in \mathbb{N}$, the family $f_\varepsilon^{(k,j)}$ uniformly satisfy standard p-growth assumptions, and $\big\{\{f_\varepsilon^{(k,j)}\}_{\varepsilon>0}\big\}_{j\in\mathbb{N}}$ is equivalent to $\{f_\varepsilon^{(k,\infty)}\}_{\varepsilon>0}$.

Assume that for all $k,j \in \mathbb{N}$ and all $U \in \mathcal{A}(\Omega)$ the Γ-limit of $\mathcal{F}_\varepsilon^{(k,j)}(\_,U)$, as $\varepsilon \to 0$, exists and is given by

$$\Gamma(L^p)\text{-}\lim_{\varepsilon\to 0} \mathcal{F}_\varepsilon^{(k,j)}(\_,U) = \mathcal{F}_0^{(k,j)}(\_,U),$$

where $\mathcal{F}_0^{(k,j)}(\_,U)$ is an integral functional with density $f_0^{(k,j)}$. Then by Theorem (2.4) there exist Carathéodory functions $f_0^{(k,\infty)} : \Omega \times \mathbb{R}^{m\times n} \to \mathbb{R}$ and the corresponding integral functionals $\mathcal{F}_0^{(k,\infty)}$ such that

$$\Gamma(L^p)\text{-}\lim_{\varepsilon\to 0} \mathcal{F}_\varepsilon^{(k,\infty)}(\_,U) = \mathcal{F}_0^{(k,\infty)}(\_,U)$$

for all $U \in \mathcal{A}(\Omega)$ and

$$\Gamma(L^p)\text{-}\lim_{j\to\infty} \mathcal{F}_0^{(k,j)}(u,U) = \lim_{j\to\infty} \mathcal{F}_0^{(k,j)}(u,U) = \mathcal{F}_0^{(k,\infty)}(u,U)$$

for all $u \in L^p(\Omega;\mathbb{R}^m)$ and $U \in \mathcal{A}(\Omega)$. From Proposition (2.14) it follows immediately that for every $j \in \mathbb{N} \cup \{\infty\}$

$$\Gamma(L^p)\text{-}\lim_{\varepsilon\to 0} \mathcal{F}_\varepsilon^{(j)}(\_,U) = \mathcal{F}_0^{(j)}(\_,U) := \inf_{k\in\mathbb{N}} \mathcal{F}_0^{(k,j)}(\_,U).$$

It remains to be proved that

$$\mathcal{F}_0^{(\infty)}(\_,U) = \lim_{j\to\infty} \mathcal{F}_0^{(j)}(\_,U) = \Gamma(L^p)\text{-}\lim_{j\to\infty} \mathcal{F}_0^{(j)}(\_,U).$$

To this end, we first infer from (2.17) that

$$\limsup_{j\to\infty} \mathcal{F}_0^{(j)}(u,U) \le \inf_{k\in\mathbb{N}} \limsup_{j\to\infty} \mathcal{F}_0^{(k,j)}(u,U) = \inf_{k\in\mathbb{N}} \mathcal{F}_0^{(k,\infty)}(u,U) = \mathcal{F}_0^{(\infty)}(u,U).$$

By (2.15) on the other hand, for any $k \in \mathbb{N}$,

$$\begin{aligned}
\liminf_{j\to\infty} \mathcal{F}_0^{(j)}(u,U) &\ge \liminf_{j\to\infty} \left(1 - \frac{1}{\alpha_U k}\right) \mathcal{F}_0^{(k,j)}(u,U) - \frac{\gamma_U}{\alpha_U k}\|u\|_{L^p(U)}^p \\
&= \left(1 - \frac{1}{\alpha_U k}\right) \mathcal{F}_0^{(k,\infty)}(u,U) - \frac{\gamma_U}{\alpha_U k}\|u\|_{L^p(U)}^p \\
&\ge \left(1 - \frac{1}{\alpha_U k}\right) \mathcal{F}_0^{(\infty)}(u,U) - \frac{\gamma_U}{\alpha_U k}\|u\|_{L^p(U)}^p,
\end{aligned}$$

and so

$$\liminf_{j\to\infty} \mathcal{F}_0^{(j)}(u,U) \ge \mathcal{F}_0^{(\infty)}(u,U).$$

Finally we note that exactly the same may be done for the Γ-$\liminf$ and Γ-$\limsup$ in place of $\liminf$, respectively, $\limsup$.

Now, if we do not a priori assume that the Γ-limits of $\mathcal{F}_\varepsilon^{(k,j)}$ exist, a diagonal sequence argument shows that for any subsequence $\{\varepsilon_i\}_{i \in \mathbb{N}}$ there is a further subsequence $\{\varepsilon_{i_l}\}_{l \in \mathbb{N}}$ such that

$$\Gamma(L^p)\text{-}\lim_{l \to \infty} \mathcal{F}_{\varepsilon_{i_l}}^{(k,j)}(\_,U) = \mathcal{F}_0^{(k,j)}(\_,U)$$

for all $k \in \mathbb{N}$, $j \in \mathbb{N} \cup \{\infty\}$ and $U \in \mathcal{A}(\Omega)$, where $\mathcal{F}_0^{(k,j)}(\_,U)$ is an integral functional with density $f_0^{(k,j)}$. Then, as shown above, $\mathcal{F}_0^{(\infty)}(u,U)$ is independent of the subsequence chosen, so that in fact

$$\Gamma(L^p)\text{-}\lim_{\varepsilon \to 0} \mathcal{F}_\varepsilon^{(\infty)}(\_,U) = \mathcal{F}_0^{(\infty)}(\_,U)$$

for $\mathcal{F}_0^{(\infty)}(u,U) = \lim_{j \to 0} \mathcal{F}_0^{(j)}(u,U)$.

Finally, the convergence of the densities now follows precisely as in the proof of Theorem (2.4). ∎

Proof (of Theorem (2.10)). This follows from the first part of the proof of Theorem (2.12) taking into account Remark (2.18). ∎

2.4   Boundary values and compactness

In this section we will first prove that the Γ-closure theorem for Gårding type functionals remains true for functionals with prescribed boundary values. On the other hand, Gårding type functionals may lack coercivity so that bounded energy sequences do not necessarily admit convergent subsequences. We will see, however, that this deficiency of compactness may be circumvented on suitable domains by imposing boundary values.

Let us for this subsection fix $\Omega \in \mathcal{A}(\mathbb{R}^n)$ and $u_0 \in W^{1,p}(\Omega;\mathbb{R}^m)$. For an integral functional $\mathcal{F}$ we denote by

$$\bar{\mathcal{F}}(u) := \begin{cases} \mathcal{F}(u), & \text{if } u \in u_0 + W_0^{1,p}(\Omega;\mathbb{R}^m), \\ \infty, & \text{otherwise,} \end{cases}$$

its restriction to prescribed boundary values u_0 on $\partial\Omega$.

Theorem (2.19). *Suppose that the family of functionals $\mathcal{F}_\varepsilon^{(j)}$, $j \in \mathbb{N} \cup \{\infty\}$, $\varepsilon > 0$, with densities $f_\varepsilon^{(j)} : \Omega \times \mathbb{R}^{m \times n} \to \mathbb{R}$ is of uniform p-Gårding type on Ω. Moreover, let*

- *$\Gamma(L^p)\text{-}\lim_{\varepsilon \to 0} \mathcal{F}_\varepsilon^{(j)} =: \mathcal{F}_0^{(j)}$ exist for each $j < \infty$,*

- *the families $\left\{\{f_\varepsilon^{(j)}\}_{\varepsilon > 0}\right\}_{j \in \mathbb{N}}$ and $\{f_\varepsilon^{(\infty)}\}_{\varepsilon > 0}$ be equivalent on Ω.*

Then also $\Gamma(L^p)\text{-}\lim_{\varepsilon \to 0} \mathcal{F}_\varepsilon^{(\infty)} =: \mathcal{F}_0^{(\infty)}$ exists,

$$\Gamma(L^p)\text{-}\lim_{\varepsilon \to 0} \bar{\mathcal{F}}_\varepsilon^{(j)} =: \bar{\mathcal{F}}_0^{(j)} \quad \forall j \in \mathbb{N} \cup \{\infty\} \quad and \quad \bar{\mathcal{F}}_0^{(\infty)} = \lim_{j \to \infty} \bar{\mathcal{F}}_0^{(j)} = \Gamma(L^p)\text{-}\lim_{j \to \infty} \bar{\mathcal{F}}_0^{(j)}.$$

We first show that prescribing boundary conditions is compatible with taking Γ-limits for a single functional satisfying a Gårding type inequality.

Lemma (2.20). *Suppose $f_\varepsilon : \Omega \times \mathbb{R}^{m \times n} \to \mathbb{R}$ are Borel functions such that*

- *for some $\beta > 0$*
$$-\beta \leq f_\varepsilon(x, X) \leq \beta(|X|^p + 1)$$
for a.e. $x \in \Omega$, all $X \in \mathbb{R}^{m \times n}$ and all $\varepsilon > 0$,

- *there are $\alpha, \gamma > 0$ such that*
$$\mathcal{F}_\varepsilon(u) \geq \alpha \int_\Omega |\nabla u(x)|^p \, dx - \gamma \int_\Omega |u(x)|^p \, dx$$
for all $u \in W^{1,p}(\Omega; \mathbb{R}^m)$ and $\varepsilon > 0$.

Then $\Gamma(L^p)\text{-}\lim_{\varepsilon \to 0} \mathcal{F}_\varepsilon = \mathcal{F}_0$ implies $\Gamma(L^p)\text{-}\lim_{\varepsilon \to 0} \bar{\mathcal{F}}_\varepsilon = \bar{\mathcal{F}}_0$.

Proof. First we show the lim inf-inequality. Suppose $u_\varepsilon \to u$ in $L^p(\Omega; \mathbb{R}^m)$.

- Suppose $\liminf_{\varepsilon \to 0} \bar{\mathcal{F}}_\varepsilon(u_\varepsilon) < \infty$. By Gårding's inequality there exists a subsequence $\{u_{\varepsilon_k}\}_{k \in \mathbb{N}} \subset u_0 + W_0^{1,p}(\Omega; \mathbb{R}^m)$ with $u_{\varepsilon_k} \rightharpoonup u$ in $W^{1,p}(\Omega; \mathbb{R}^m)$. Then also $u \in u_0 + W_0^{1,p}(\Omega; \mathbb{R}^m)$, and therefore $\bar{\mathcal{F}}_0(u) < \infty$. Hence, $\bar{\mathcal{F}}_0(u) = \infty$ implies $\liminf_{\varepsilon \to 0} \bar{\mathcal{F}}_\varepsilon(u_\varepsilon) = \infty$.

- If $\bar{\mathcal{F}}_0(u) < \infty$, then
$$\liminf_{\varepsilon \to 0} \bar{\mathcal{F}}_\varepsilon(u_\varepsilon) \geq \liminf_{\varepsilon \to 0} \mathcal{F}_\varepsilon(u_\varepsilon) \geq \mathcal{F}_0(u) = \bar{\mathcal{F}}_0(u).$$

Now we prove the existence of the recovery sequence. Consider an arbitrary subsequence $\varepsilon_j \searrow 0$. According to the pointwise Urysohn property from Theorem (B.3), for given $u \in L^p(\Omega; \mathbb{R}^m)$ we only have to provide a recovery sequence $v_k \to u$ in $L^p(\Omega; \mathbb{R}^m)$ along a suitable subsequence $\{\varepsilon_{j_k}\}_{k \in \mathbb{N}}$ of $\{\varepsilon_j\}_{j \in \mathbb{N}}$.

Let u_ε be a recovery sequence for the original functional $\mathcal{F}_0$ for u, i.e.,
$$u_\varepsilon \to u \quad \text{in } L^p(\Omega; \mathbb{R}^m) \quad \text{and} \quad \lim_{\varepsilon \to 0} \mathcal{F}_\varepsilon(u_\varepsilon) = \mathcal{F}_0(u).$$

If $\mathcal{F}_0(u) = \infty$, then also $\lim_{\varepsilon \to 0} \bar{\mathcal{F}}_\varepsilon(u_\varepsilon) = \bar{\mathcal{F}}_0(u)$ and the claim follows. If $\mathcal{F}_0(u) < \infty$, then by Gårding's inequality $\{u_\varepsilon\}_{\varepsilon > 0}$ is bounded in $W^{1,p}(\Omega; \mathbb{R}^m)$. Employing Lemma (C.4), we find a subsequence $\{\varepsilon_{j_k}\}_{k \in \mathbb{N}}$ and $v_k \in u_0 + W_0^{1,p}(\Omega; \mathbb{R}^m)$ such that $\{|\nabla v_k|^p\}_{k \in \mathbb{N}}$ is equiintegrable in Ω,
$$|A_k| \to 0 \quad \text{for} \quad A_k := \{u_{\varepsilon_{j_k}} \neq v_k \text{ or } \nabla u_{\varepsilon_{j_k}} \neq \nabla v_k\},$$
and $v_k \rightharpoonup u$ in $W^{1,p}(\Omega; \mathbb{R}^m)$. But then
$$\begin{aligned}
\bar{\mathcal{F}}_{\varepsilon_{j_k}}(v_k) &= \mathcal{F}_{\varepsilon_{j_k}}(v_k) \\
&= \int_{\Omega \setminus A_k} f_{\varepsilon_{j_k}}(x, \nabla u_{\varepsilon_{j_k}}(x)) \, dx + \int_{A_k} f_{\varepsilon_{j_k}}(x, \nabla v_k(x)) \, dx \\
&\leq \mathcal{F}_{\varepsilon_{j_k}}(u_{\varepsilon_{j_k}}) + \beta|A_k| + \int_{A_k} \beta(1 + |\nabla v_k(x)|^p) \, dx,
\end{aligned}$$
and thus
$$\limsup_{k \to \infty} \bar{\mathcal{F}}_{\varepsilon_{j_k}}(v_k) \leq \limsup_{k \to \infty} \mathcal{F}_{\varepsilon_{j_k}}(u_{\varepsilon_{j_k}}) \leq \mathcal{F}_0(u) = \bar{\mathcal{F}}_0(u). \qquad \blacksquare$$

Proof (of Theorem (2.19)). The claim $\Gamma(L^p)\text{-}\lim_{\varepsilon \to 0} \mathcal{F}_\varepsilon^{(\infty)} =: \mathcal{F}_0^{(\infty)}$ was shown in Theorem (2.10). The remaining assertions follow from the same theorem and Lemma (2.20) by noting that also the family $\mathcal{F}_0^{(j)}$, $j \in \mathbb{N} \cup \{\infty\}$, is of uniform p-Gårding type on Ω with the same constants α_Ω and γ_Ω:

Firstly, the existence of densities follows from Proposition (2.14) with Remark (2.13). Secondly, for given j let $\{u_\varepsilon\}_\varepsilon$ be a recovery sequence for $u \in W^{1,p}(\Omega; \mathbb{R}^m)$. By Gårding's inequality and boundedness of $\{\mathcal{F}_\varepsilon^{(j)}(u_\varepsilon)\}_\varepsilon$, we have $u_\varepsilon \rightharpoonup u$ in $W^{1,p}(\Omega; \mathbb{R}^m)$, and so

$$
\begin{aligned}
\mathcal{F}_0^{(j)}(u) &= \lim_{\varepsilon \to 0} \mathcal{F}_\varepsilon^{(j)}(u_\varepsilon) \\
&\geq \liminf_{\varepsilon \to 0} \left(\alpha_\Omega \|\nabla u_\varepsilon\|_{L^p(\Omega)}^p - \gamma_\Omega \|u_\varepsilon\|_{L^p(\Omega)}^p \right) \\
&\geq \alpha_\Omega \|\nabla u\|_{L^p(\Omega)}^p - \gamma_\Omega \|u\|_{L^p(\Omega)}^p. \quad \blacksquare
\end{aligned}
$$

In view of our application to homogenization theory to be discussed below, we observe that Poincaré's inequality guarantees coercivity on sufficiently small domains.

Proposition (2.21). *Suppose $f : \Omega \times \mathbb{R}^{m \times n} \to \mathbb{R}$ is a Borel function such that*

- *for some $\beta > 0$ and $1 < p < \infty$*

$$
-\beta \leq f(x, X) \leq \beta(|X|^p + 1)
$$

  for all $x \in \Omega$ and $X \in \mathbb{R}^{m \times n}$,

- *there are $\alpha_\Omega, \gamma_\Omega > 0$ such that*

$$
\mathcal{F}(u, \Omega) \geq \alpha_\Omega \int_\Omega |\nabla u(x)|^p \, dx - \gamma_\Omega \int_\Omega |u(x)|^p \, dx
$$

  for all $u \in W^{1,p}(\Omega; \mathbb{R}^m)$.

Then, if $U \subset \Omega$ is sufficiently small, there are constants $A > 0$ and B such that

$$
\mathcal{F}(u, U) \geq A \int_U |\nabla u(x)|^p \, dx - B
$$

for all $u \in u_0 + W_0^{1,p}(U; \mathbb{R}^m)$.

Proof. Take any $U \in \mathcal{A}(\Omega)$. For $u \in u_0 + W_0^{1,p}(U; \mathbb{R}^m)$ let $\bar{u} \in W^{1,p}(\Omega; \mathbb{R}^m)$ be its extension by u_0 on $\Omega \setminus U$. We have

$$
\begin{aligned}
\mathcal{F}&(u, U) \\
&= \mathcal{F}(\bar{u}, \Omega) - \int_{\Omega \setminus U} f(x, \nabla u_0(x)) \, dx \\
&\geq \int_\Omega \left(\alpha_\Omega |\nabla \bar{u}(x))|^p - \gamma_\Omega |\bar{u}(x)|^p \right) dx - \int_{\Omega \setminus U} \left(\beta |\nabla u_0(x)|^p + \beta \right) dx \\
&= \int_U \left(\alpha_\Omega |\nabla u(x))|^p - \gamma_\Omega |u(x)|^p \right) dx - \int_{\Omega \setminus U} \left((\beta - \alpha_\Omega)|\nabla u_0(x)|^p + \gamma_\Omega |u_0(x)|^p + \beta \right) dx.
\end{aligned}
$$

Now, if C_P denotes the Poincaré constant of U, then

$$
\begin{aligned}
\int_U |u(x)|^p\,dx &\leq \int_U \left(2^{p-1}|u(x)-u_0(x)|^p + 2^{p-1}|u_0(x)|^p\right)\,dx \\
&\leq 2^{p-1}C_\mathrm{P}^p \int_U |\nabla u(x)-\nabla u_0(x)|^p\,dx + 2^{p-1}\int_U |u_0(x)|^p\,dx \\
&\leq 4^{p-1}C_\mathrm{P}^p \int_U |\nabla u(x)|^p\,dx + 4^{p-1}C_\mathrm{P}^p \int_U |\nabla u_0(x)|^p\,dx + 2^{p-1}\int_U |u_0(x)|^p\,dx.
\end{aligned}
$$

The Poincaré constant depends only on the width of U. The assertion follows if we choose U so narrow that $A := \alpha_\Omega - 4^{p-1}C_\mathrm{P}^p\gamma_\Omega$ is positive. ∎

The proof shows that A and B only depend on the width of U (through its Poincaré constant), $p, \alpha_\Omega, \gamma_\Omega, \beta$ and $\|u_0\|_{W^{1,p}(\Omega)}$.

2.5   A perturbation and a relaxation result

When applied to j- or ε-independent families, our Γ-closure theorems immediately imply the following perturbation and relaxation results. We only consider their formulation on a single domain Ω; the adaptation to variable domains is straightforward.

The first easy consequence of the Γ-closure theorems is a stability result for Γ-limits under equivalent perturbations of the densities.

Theorem (2.22). *Let $\Omega \subset \mathbb{R}^n$ be bounded and open. Suppose that the families of functionals $\mathcal{F}_\varepsilon$ and $\mathcal{G}_\varepsilon$, $\varepsilon > 0$, with densities $f_\varepsilon, g_\varepsilon : \Omega \times \mathbb{R}^{m\times n} \to \mathbb{R}$, respectively, are of uniform p-Gårding type on Ω. Assume that $\Gamma(L^p)\text{-}\lim_{\varepsilon\to 0}\mathcal{F}_\varepsilon = \mathcal{F}_0$ and*

$$
\limsup_{\varepsilon\to 0}\int_\Omega \sup_{|X|\leq R} |f_\varepsilon(x,X)-g_\varepsilon(x,X)|\,dx = 0
$$

for all $R > 0$. Then also $\Gamma(L^p)\text{-}\lim_{\varepsilon\to 0}\mathcal{G}_\varepsilon = \mathcal{F}_0$.

Schematically we can recapitulate the theorem with the following diagram:

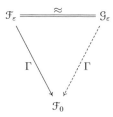

Proof. By setting $\mathcal{F}_\varepsilon^{(j)} := \mathcal{F}_\varepsilon$ for all $j \in \mathbb{N}$ and $\varepsilon \geq 0$ and $\mathcal{G}_\varepsilon := \mathcal{F}_\varepsilon^{(\infty)}$ for $\varepsilon > 0$, it follows directly from Theorem (2.10) that

$$
\Gamma(L^p)\text{-}\lim_{\varepsilon\to 0}\mathcal{G}_\varepsilon = \lim_{j\to\infty}\mathcal{F}_0 = \mathcal{F}_0. \quad \blacksquare
$$

Our second straightforward application shows that a sequence of functionals equivalent to some fixed functional Γ-converges to the relaxation of this functional.

Theorem (2.23). *Let $\Omega \subset \mathbb{R}^n$ be bounded and open. Suppose that the family of functionals $\mathcal{F}^{(j)}$, $j \in \mathbb{N} \cup \{\infty\}$, with densities $f^{(j)} : \Omega \times \mathbb{R}^{m \times n} \to \mathbb{R}$ are of uniform p-Gårding type on Ω. Assume that*

$$\lim_{j \to \infty} \int_\Omega \sup_{|X| \leq R} |f^{(j)}(x, X) - f^{(\infty)}(x, X)| \, dx = 0$$

for all $R > 0$. Then $\Gamma(L^p)$-$\lim_{j \to \infty} \mathcal{F}^{(j)} = \operatorname{lsc} \mathcal{F}^{(\infty)}$.

Here, the result can be presented by

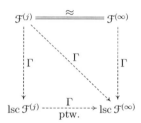

Proof. Let $\mathcal{F}_\varepsilon^{(j)} := \mathcal{F}^{(j)}$ for all $j \in \mathbb{N} \cup \{\infty\}$ and $\varepsilon > 0$. Since these sequences are constant in ε,

$$\Gamma(L^p)\text{-}\lim_{\varepsilon \to 0} \mathcal{F}_\varepsilon^{(j)} = \operatorname{lsc} \mathcal{F}^{(j)}.$$

Hence, we may apply Theorem (2.10), which yields

$$\Gamma(L^p)\text{-}\lim_{j \to \infty} \operatorname{lsc} \mathcal{F}^{(j)} = \lim_{j \to \infty} \operatorname{lsc} \mathcal{F}^{(j)} = \operatorname{lsc} \mathcal{F}^{(\infty)}.$$

Noting that $\Gamma(L^p)$-$\lim_{j \to \infty} \mathcal{F}^{(j)} = \Gamma(L^p)$-$\lim_{j \to \infty} \operatorname{lsc} \mathcal{F}^{(j)}$ finishes the proof. ∎

The lower semicontinuous envelope appearing in Theorem (2.23) is still given by the quasiconvex envelope as in the case of standard growth:

Corollary (2.24). *Let $\Omega \subset \mathbb{R}^n$ be a bounded open set. If $\mathcal{F}$ is a functional of p-Gårding type on Ω with a Carathéodory density $f : \Omega \times \mathbb{R}^{m \times n} \to \mathbb{R}$, then its lower semicontinuous envelope in $L^p(\Omega)$ is*

$$\operatorname{lsc} \mathcal{F}(u) = \begin{cases} \int_\Omega f^{qc}(x, \nabla u(x)) \, dx, & u \in W^{1,p}(\Omega; \mathbb{R}^n), \\ \infty, & else. \end{cases}$$

Proof. We apply Proposition (2.14) to the constant family given by f. Since each $f^{(k)}(x, X) := f(x, X) + \frac{1}{k}|X|^p$ has a standard p-growth, by Theorem (B.5) the lower semicontinuous envelope $\operatorname{lsc} \mathcal{F}^{(k)}$ has density $(f^{(k)})^{qc}$. The result follows as $(f^{(k)})^{qc} \searrow f^{qc}$ by formula (A.2) and the dominated convergence theorem. ∎

2.6 Commutability of Γ-limits

In the general case with a doubly indexed family of functionals considered in the closure theorems above, the natural question arises: Do the Γ-limits as $\varepsilon \to 0$ and as $j \to \infty$ commute, and is the limiting functional some simultaneous limit of $\mathcal{F}^{(j_k)}_{\varepsilon_k}$ as $k \to \infty$? In general, this is not the case as the example

$$f^{(j)}_{\varepsilon}(x, X) := \begin{cases} |X|^p, & \text{if } \varepsilon < \frac{1}{j}, \\ 2|X|^p, & \text{if } \varepsilon \geq \frac{1}{j}, \end{cases} \quad \text{for } j \in \mathbb{N}, \qquad f^{(\infty)}_{\varepsilon}(x, X) := |X|^p \quad \forall \varepsilon$$

shows.

However, again as a direct application of our closure results, we obtain that a stronger notion of equivalence does in fact imply commutability of these Γ-limits. This stronger condition in particular is satisfied if

$$\lim_{j \to \infty} \sup_{\varepsilon > 0} \int_U \sup_{|X| \leq R} |f^{(j)}_{\varepsilon}(x, X) - f^{(\infty)}_{\varepsilon}(x, X)| \, dx = 0$$

and thus holds true in our applications to homogenization to be discussed in the next section. In the following it suffices to consider a single domain Ω.

Theorem (2.25). *Let $\Omega \subset \mathbb{R}^n$ be bounded and open. Suppose that the family of functionals $\mathcal{F}^{(j)}_{\varepsilon}$, $j \in \mathbb{N} \cup \{\infty\}$, $\varepsilon > 0$, with densities $f^{(j)}_{\varepsilon} : \Omega \times \mathbb{R}^{m \times n} \to \mathbb{R}$ is of uniform p-Gårding type on Ω. Assume that*

- *for each $j < \infty$ the Γ-limit $\Gamma(L^p)\text{-}\lim_{\varepsilon \to 0} \mathcal{F}^{(j)}_{\varepsilon} =: \mathcal{F}^{(j)}_0$ exists,*

- *the families $\{\{f^{(j)}_{\varepsilon}\}_{\varepsilon > 0}\}_{j \in \mathbb{N}}$ and $\{f^{(\infty)}_{\varepsilon}\}_{\varepsilon > 0}$ are equivalent on Ω and moreover*

$$\lim_{j \to \infty} \int_\Omega \sup_{|X| \leq R} |f^{(j)}_{\varepsilon}(x, X) - f^{(\infty)}_{\varepsilon}(x, X)| \, dx = 0$$

for every $R > 0$ and all $\varepsilon > 0$.

Then

$$\Gamma(L^p)\text{-}\lim_{\varepsilon \to 0} \mathcal{F}^{(\infty)}_{\varepsilon} = \Gamma(L^p)\text{-}\lim_{j \to \infty} \mathcal{F}^{(j)}_0 =: \mathcal{F}^{(\infty)}_0.$$

Moreover, $\Gamma(L^p)\text{-}\lim_{j \to \infty} \mathcal{F}^{(j)}_{\varepsilon} = \text{lsc } \mathcal{F}^{(\infty)}_{\varepsilon}$ and

$$\Gamma(L^p)\text{-}\lim_{\varepsilon \to 0} \left(\Gamma(L^p)\text{-}\lim_{j \to \infty} \mathcal{F}^{(j)}_{\varepsilon} \right) = \Gamma(L^p)\text{-}\lim_{j \to \infty} \left(\Gamma(L^p)\text{-}\lim_{\varepsilon \to 0} \mathcal{F}^{(j)}_{\varepsilon} \right),$$

i.e., the following diagram commutes:

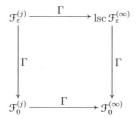

Proof. The first assertion follows immediately from Theorem (2.10). Moreover, Theorem (2.23) applied to $\mathcal{F}_\varepsilon^{(j)}$ with fixed ε gives

$$\Gamma(L^p)\text{-}\lim_{j\to\infty} \mathcal{F}_\varepsilon^{(j)} = \operatorname{lsc} \mathcal{F}_\varepsilon^{(\infty)}.$$

It remains to note that $\Gamma(L^p)\text{-}\lim_{\varepsilon\to 0} \mathcal{F}_\varepsilon^{(\infty)} = \Gamma(L^p)\text{-}\lim_{\varepsilon\to 0} \operatorname{lsc} \mathcal{F}_\varepsilon^{(\infty)}$. ∎

The diagram below presents the assumptions (solid lines) and consequences (dashed) in Theorem (2.25). The arrows indicate Γ-convergence with respect to j, ε or of the constant sequence with fixed indices (the relaxation).

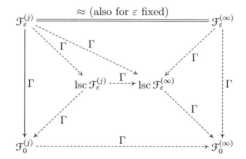

In fact, under the condition

$$\lim_{j\to\infty}\sup_{\varepsilon>0} \int_\Omega \sup_{|X|\le R} |f_\varepsilon^{(j)}(x,X) - f_\varepsilon^{(\infty)}(x,X)|\, dx = 0$$

we also have Γ-convergence along any diagonal sequence $\{(\varepsilon_k, j_k)\}_{k\in\mathbb{N}}$.

Theorem (2.26). *Let $\Omega \subset \mathbb{R}^n$ be bounded and open. Suppose that the family of functionals $\mathcal{F}_\varepsilon^{(j)}$, $j \in \mathbb{N}\cup\{\infty\}$, $\varepsilon > 0$, with densities $f_\varepsilon^{(j)} : \Omega \times \mathbb{R}^{m\times n} \to \mathbb{R}$ is of uniform p-Gårding type on Ω. Let $\{j_k\}_{k\in\mathbb{N}}$ and $\{\varepsilon_k\}_{k\in\mathbb{N}}$ be subsequences of $\mathbb{N}$ and ε, respectively. Assume that $\Gamma(L^p)\text{-}\lim_{\varepsilon\to 0}\mathcal{F}_\varepsilon^{(\infty)} = \mathcal{F}_0^{(\infty)}$ and*

$$\lim_{k\to\infty} \int_\Omega \sup_{|X|\le R} |f_{\varepsilon_k}^{(j_k)}(x,X) - f_{\varepsilon_k}^{(\infty)}(x,X)|\, dx = 0$$

for every $R > 0$ and all $\varepsilon > 0$. Then $\Gamma(L^p)\text{-}\lim_{k\to\infty}\mathcal{F}_{\varepsilon_k}^{(j_k)} = \mathcal{F}_0^{(\infty)}$.

Proof. This is nothing but Theorem (2.22) applied to $\mathcal{F}_{\varepsilon_k}^{(j_k)}$ and $\mathcal{F}_{\varepsilon_k}^{(\infty)}$. ∎

Note that $\Gamma(L^p)\text{-}\lim_{\varepsilon\to 0}\mathcal{F}_\varepsilon^{(\infty)} = \mathcal{F}_0^{(\infty)}$ is known to exist under the assumptions of Theorem (2.12).

2.7 Homogenization of Gårding type functionals

As a corollary to our Γ-closure and commutability theorems, we first obtain a homogenization closure theorem. It generalizes the corresponding result of Braides to integral densities that are not assumed to be quasiconvex in their second argument and that instead of a standard p-growth assumption are only assumed to induce integral functionals of Gårding type. Secondly, we also obtain a general criterion for the interchangeability of homogenizing and taking the Γ-limit of a sequence of functionals. In view of our applications in the following sections, we state the closure theorem on variable domains and with densities bounded from below by appropriate family inducing Gårding's inequality.

Definition (2.27). Let a Borel function $f : \mathbb{R}^n \times \mathbb{R}^{m \times n} \to \mathbb{R}$ for some $\beta > 0$ fulfil

$$-\beta \leq f(x, X) \leq \beta(|X|^p + 1)$$

for a.e. $x \in \mathbb{R}^n$ and all $X \in \mathbb{R}^{m \times n}$. We say that f is *homogenizable* if there exists a Borel function $f_{\mathrm{hom}} : \mathbb{R}^{m \times n} \to \mathbb{R}$ such that

$$\mathcal{F}_{\mathrm{hom}}(\_, U) = \Gamma(L^p)\text{-}\lim_{\varepsilon \to 0} \mathcal{F}_\varepsilon(\_, U)$$

for every $U \in \mathcal{A}(\mathbb{R}^n)$ where for $u \in W^{1,p}(\mathbb{R}^n; \mathbb{R}^m)$ the functionals are defined as

$$\mathcal{F}_\varepsilon(u, U) := \int_U f(\tfrac{x}{\varepsilon}, \nabla u(x)) \, dx \quad \text{and} \quad \mathcal{F}_{\mathrm{hom}}(u, U) := \int_U f_{\mathrm{hom}}(\nabla u(x)) \, dx$$

and take value ∞ everywhere else on $L^p(\mathbb{R}^n; \mathbb{R}^m)$.

We introduce a type of lower bounds related to Gårding's inequality in order to facilitate the formulation of the subsequent claims. Namely, since ε appears in a special way, it is more convenient to have a growth condition on the function f itself rather than for every ε on $(x, X) \mapsto f(\tfrac{x}{\varepsilon}, X)$. However, should such a lower bound be too restrictive, one has to check if at least the lower bound in the form from the previous sections is fulfilled.

Definition (2.28).

(a) A Borel function $g : \mathbb{R}^{m \times n} \to \mathbb{R}$ is a *lower bound of p-Gårding type* if for every $U \in \mathcal{A}(\mathbb{R}^n)$ there exist $\alpha_U > 0$ and $\gamma_U \in \mathbb{R}$ such that

$$\int_U g(\nabla u(x)) \, dx \geq \alpha_U \int_U |\nabla u(x)|^p \, dx - \gamma_U \int_U |u(x)|^p \, dx$$

for all $u \in W^{1,p}(U; \mathbb{R}^m)$.

(b) Lower bounds of p-Gårding type $g^{(j)}$ make a *family of lower bounds of uniform p-Gårding type* if for every $U \in \mathcal{A}(\mathbb{R}^n)$ the constants α_U, γ_U may be chosen uniformly in j.

If f is $\mathbb{I}^n$-periodic in the first variable and has a standard p-growth, then it is homogenizable, and f_{hom} is given by the formula, stated in Section 1.4. However, the same representation formula holds also for any homogenizable function of standard growth, and in fact the assumption on the lower bound can also be relaxed. That is a consequence of Theorem (2.19) and Proposition (2.21).

Proposition (2.29). *Suppose that for a Borel function* $f : \mathbb{R}^n \times \mathbb{R}^{m \times n} \to \mathbb{R}$ *there exist* $\beta > 0$ *and a lower bound of p-Gårding type g such that*

$$\max\{-\beta, g(X)\} \leq f(x, X) \leq \beta(|X|^p + 1)$$

for a.e. $x \in \mathbb{R}^n$ *and all* $X \in \mathbb{R}^{m \times n}$. *If f is homogenizable, then* f_{hom} *is a quasiconvex function, and for every* $X \in \mathbb{R}^{m \times n}$

$$f_{\mathrm{hom}}(X) = \lim_{t \to \infty} \inf \left\{ \frac{1}{t^n |U|} \int_{tU} f(x, X + \nabla\varphi(x))\; dx : \varphi \in W_0^{1,p}(tU; \mathbb{R}^m) \right\}$$

for any $U \in \mathcal{A}(\mathbb{R}^n)$.

Proof. Quasiconvexity of f_{hom} was actually shown in the proof of Proposition (2.14). Fix any $U \in \mathcal{A}(\mathbb{R}^n)$ and any sequence $t_k \nearrow \infty$. Let Ω be an open ball centred at 0 such that $U \subset \Omega$. Take $X \in \mathbb{R}^{m \times n}$, and set $u_0(x) := \ell_X(x) = Xx$. By Proposition (2.21) the functionals $\bar{\mathcal{F}}_\varepsilon(\_, U_M)$ are equicoercive on $W^{1,p}(U_M; \mathbb{R}^m)$ for $U_M := \frac{1}{M}U$ with $M > 0$ sufficiently large. Quasiconvexity of f_{hom}, Lemma (2.20) and Theorem (B.2) thus imply that for $\varepsilon_k := \frac{1}{t_k M}$

$$
\begin{aligned}
f_{\mathrm{hom}}(X) &= \min \left\{ \frac{1}{|U_M|} \mathcal{F}_{\mathrm{hom}}(u, U_M) : u \in \ell_X + W_0^{1,p}(U_M; \mathbb{R}^m) \right\} \\
&= \min \left\{ \frac{1}{|U_M|} \bar{\mathcal{F}}_{\mathrm{hom}}(u, U_M) : u \in L^p(U_M; \mathbb{R}^m) \right\} \\
&= \lim_{k \to \infty} \inf \left\{ \frac{1}{|U_M|} \bar{\mathcal{F}}_{\varepsilon_k}(u, U_M) : u \in L^p(U_M; \mathbb{R}^m) \right\} \\
&= \lim_{k \to \infty} \inf \left\{ \frac{1}{|U_M|} \int_{U_M} f(t_k M x, \nabla u(x))\; dx : u \in \ell_X + W_0^{1,p}(U_M; \mathbb{R}^m) \right\} \\
&= \lim_{k \to \infty} \inf \left\{ \frac{1}{t_k^n |U|} \int_{t_k U} f(x, X + \nabla\varphi(x))\; dx : \varphi \in W_0^{1,p}(t_k U; \mathbb{R}^m) \right\}. \quad \blacksquare
\end{aligned}
$$

Remark (2.30). If the function f in Proposition (2.29) is Carathéodory, then f^{qc} is homogenizable as well, and $(f^{\mathrm{qc}})_{\mathrm{hom}} = f_{\mathrm{hom}}$. This is an immediate consequence of Corollary (2.24).

Now we are in the position to state an analogue to the homogenization closure theorem (3.5. Corollary in [Br:86]).

Theorem (2.31). *Let for a family of Borel functions*

$$f^{(j)} : \mathbb{R}^n \times \mathbb{R}^{m \times n} \to \mathbb{R}, \quad j \in \mathbb{N} \cup \{\infty\}$$

exist a constant $\beta > 0$ *and a family of lower bounds of uniform p-Gårding type* $g^{(j)}$ *with*

$$\max\{-\beta, g^{(j)}(X)\} \leq f^{(j)}(x, X) \leq \beta(|X|^p + 1)$$

for a.e. $x \in \mathbb{R}^n$, *every* $X \in \mathbb{R}^{m \times n}$ *and every* $j \in \mathbb{N} \cup \{\infty\}$. *Suppose that*

- *for every finite j the function* $f^{(j)}$ *is homogenizable,*

- *for every $R > 0$*

$$\lim_{j \to \infty} \limsup_{T \to \infty} \frac{1}{(2T)^n} \int_{(-T,T)^n} \sup_{|X| \leq R} |f^{(j)}(x, X) - f^{(\infty)}(x, X)| \, dx = 0.$$

Then also $f^{(\infty)}$ is homogenizable, and the limiting density $f_{\mathrm{hom}}^{(\infty)}$ is given by

$$f_{\mathrm{hom}}^{(\infty)}(X) = \lim_{k \to \infty} \inf \left\{ \frac{1}{k^n} \int_{k\mathbb{I}^n} f^{(\infty)}(x, X + \nabla \varphi(x)) \, dx : \varphi \in W_0^{1,p}(k\mathbb{I}^n; \mathbb{R}^m) \right\}$$

and satisfies $f_{\mathrm{hom}}^{(\infty)}(X) = \lim_{j \to \infty} f_{\mathrm{hom}}^{(j)}(X)$ for every $X \in \mathbb{R}^{m \times n}$.

Remark (2.32). We allowed the lower bounds to depend on j in order to include relevant examples from elasticity theory, as we will see in Lemma (3.7).

Except for the representation formula of $f_{\mathrm{hom}}^{(\infty)}$ proved above, this is a direct consequence of Theorem (2.12).

Proof. Let

$$f_\varepsilon^{(j)}(x, X) := f^{(j)}(\tfrac{x}{\varepsilon}, X)$$

for $j \in \mathbb{N} \cup \{\infty\}$ and $\varepsilon > 0$, and set

$$f_0^{(j)}(x, X) := f_{\mathrm{hom}}^{(j)}(X)$$

for $j \in \mathbb{N}$. For $U \in \mathcal{A}(\mathbb{R}^n)$ with $U \subset (-T_0, T_0)^n$ and $R > 0$, we then have

$$\lim_{j \to \infty} \limsup_{\varepsilon \to 0} \int_U \sup_{|X| \leq R} |f_\varepsilon^{(j)}(x, X) - f_\varepsilon^{(\infty)}(x, X)| \, dx$$

$$\leq \limsup_{j \to \infty} \limsup_{\varepsilon \to 0} \int_{(-T_0, T_0)^n} \sup_{|X| \leq R} |f^{(j)}(\tfrac{x}{\varepsilon}, X) - f^{(\infty)}(\tfrac{x}{\varepsilon}, X)| \, dx$$

$$= \limsup_{j \to \infty} \limsup_{\varepsilon \to 0} \varepsilon^n \int_{(-T_0/\varepsilon, T_0/\varepsilon)^n} \sup_{|X| \leq R} |f^{(j)}(x, X) - f^{(\infty)}(x, X)| \, dx$$

$$= \limsup_{j \to \infty} \limsup_{T \to \infty} \frac{(2T_0)^n}{(2T)^n} \int_{(-T,T)^n} \sup_{|X| \leq R} |f^{(j)}(x, X) - f^{(\infty)}(x, X)| \, dx$$

$$= 0.$$

Thus we may apply Theorem (2.12), and deduce that there exists a Carathéodory function $f_0^{(\infty)} : \mathbb{R}^n \times \mathbb{R}^{m \times n} \to \mathbb{R}$ and the corresponding integral functional $\mathcal{F}_0^{(\infty)}$ such that

$$\Gamma(L^p)\text{-} \lim_{\varepsilon \to 0} \mathcal{F}_\varepsilon^{(\infty)}(\_, U) = \mathcal{F}_0^{(\infty)}(\_, U)$$

for all $U \in \mathcal{A}(\mathbb{R}^n)$, and moreover

$$f_0^{(j)}(\_, X) \overset{*}{\rightharpoonup} f_0^{(\infty)}(\_, X) \quad \text{in } L^\infty(\mathbb{R}^n)$$

for all $X \in \mathbb{R}^{m \times n}$. Since $f_0^{(j)}(x, X) = f_{\mathrm{hom}}^{(j)}(X)$ is independent of x, we in fact have that $f_0^{(\infty)}(x, X) = f_{\mathrm{hom}}^{(\infty)}(X)$ for some continuous function $f_{\mathrm{hom}}^{(\infty)}$ for a.e. $x \in \mathbb{R}^n$ and all $X \in \mathbb{R}^{m \times n}$. In particular,

$$\lim_{j \to \infty} f_{\mathrm{hom}}^{(j)}(X) = f_{\mathrm{hom}}^{(\infty)}(X)$$

for all $X \in \mathbb{R}^{m \times n}$. The representation formula now follows from Proposition (2.29). ∎

Let us for the sake of completeness state the following extension of the periodic homogenization result to functionals of Gårding type:

Corollary (2.33). *If a Borel function $f : \mathbb{R}^n \times \mathbb{R}^{m \times n} \to \mathbb{R}$*

- *is $\mathbb{I}^n$-periodic in the first variable,*
- *has the growth described in Proposition (2.29),*

then f is homogenizable.

Proof. Define $f^{(j)}(x, X) := f(x, X) + \frac{1}{j}|X|^p$, and apply Theorem (2.31). ■

We finally give an equivalence condition adapted to homogenization leading to a Γ-commuting diagram as in Theorem (2.25). The analogous adaptation of Theorem (2.26) is straightforward.

Theorem (2.34). *Suppose $f^{(j)}$, $j \in \mathbb{N} \cup \{\infty\}$ satisfy the assumptions of Theorem (2.31) and in addition*

$$\lim_{j \to \infty} \int_{(-T,T)^n} \sup_{|X| \leq R} |f^{(j)}(x, X) - f^{(\infty)}(x, X)| \, dx = 0$$

for all $R, T > 0$. Then the limit $f^{(\infty)}$ is also homogenizable, and the following diagram commutes for every $U \in \mathcal{A}(\mathbb{R}^n)$:

$$
\begin{array}{ccc}
\mathcal{F}_\varepsilon^{(j)}(\_,U) & \xrightarrow{\;\;\Gamma\;\;} & \operatorname{lsc} \mathcal{F}_\varepsilon^{(\infty)}(\_,U) \\
\Big\downarrow{\scriptstyle\Gamma} & & \Big\downarrow{\scriptstyle\Gamma} \\
\mathcal{F}_{\hom}^{(j)}(\_,U) & \xrightarrow{\;\;\Gamma\;\;} & \mathcal{F}_{\hom}^{(\infty)}(\_,U)
\end{array}
$$

Proof. This is an immediate consequence of Theorem (2.25). ■

Chapter 3

Applications in elasticity theory

Equipped with the results from the previous chapter (hence, still $p > 1$), we return to the motivating example from elasticity theory presented in Chapter 1. Our aim was to prove Theorem (1.1). We saw in Chapter 2 that actually the diagrams may commute even if the densities are not $\mathbb{I}^n$-periodic. Our plan for this chapter is therefore as follows. We will define a fairly general class of families that possess typical symmetry and growth properties of one- or multiwell densities in (non)linear elasticity. We will check their compatibility with Theorem (2.31). Then we will restate Theorem (1.1) in an appropriately generalized form. The proof will after all the preparations consist just of bringing the pieces together and checking equivalence.

Let us highlight two additional results from this chapter. Theorem (3.5) provides an alternative form of the geometric rigidity result. Moreover, we will in Theorem (3.14) extend Zhang's result on the lower bound of the quasiconvex envelope of $\text{dist}^2(\_, \text{SO}(n))$ to any exponent $p > 1$.

3.1 Functionals for microstructured multiwell materials

In Section 1.5 we presented a concrete model with the stored-energy function having multiple wells in the neighbourhood of $\text{SO}(n)$. Let us here introduce a larger class of admissible functions that possess essential properties of reasonable mechanical models, namely, the frame-indifference and good growth conditions.

Definition (3.1). A family of Borel functions $W^{(\delta)} : \mathbb{R}^n \times \mathbb{R}^{n \times n} \to \mathbb{R}$, $\delta > 0$, is an *admissible family of stored-energy functions for nonlinear elasticity* if for every $\delta > 0$, a.e. $x \in \mathbb{R}^n$, all $X \in \mathbb{R}^{n \times n}$, all $R \in \text{SO}(n)$ and some $\alpha, \beta > 0$

- $W^{(\delta)}(x, RX) = W^{(\delta)}(x, X)$,

- $\alpha \, \text{dist}^p\left(X, \text{SO}(n)\right) - \beta \, \delta^p \leq W^{(\delta)}(x, X) \leq \beta\left(|X - I|^p + \delta^p\right)$.

For any positive matrix $S \in \mathbb{R}^{n \times n}$ and any rotation $R \in \text{SO}(n)$, we have

$$|X - R| = |X - R(I + \delta S) + \delta RS| \leq |X - R(I + \delta S)| + \delta|S|.$$

Therefore,

$$\text{dist}^p\big(X, \text{SO}(n)(I + \delta S)\big) \geq \tfrac{1}{2^{p-1}}\text{dist}^p(X, \text{SO}(n)) - \delta^p |S|^p.$$

Similarly,

$$\text{dist}^p\big(X, \text{SO}(n)(I + \delta S)\big) \leq 2^{p-1}\Big(\text{dist}^p\big(X, \text{SO}(n)\big) + \delta^p |S|^p\Big).$$

Thus, the growth conditions imposed in Theorem (1.1) are compatible with those in the definition above. We also need an analogous notion for linear elasticity. Frame-indifference in the nonlinear case corresponds to dependence only on the symmetric part of a matrix. Moreover, we adjust the growth assumption.

Definition (3.2). A Borel function $V : \mathbb{R}^n \times \mathbb{R}^{n\times n} \to \mathbb{R}$ is an *admissible stored-energy function for linear elasticity* if for a.e. $x \in \mathbb{R}^n$, all $X \in \mathbb{R}^{n\times n}$ and suitable $\alpha, \beta > 0$

- $V(x, X) = V(x, X_{\text{sym}})$,

- $\alpha |X_{\text{sym}}|^p - \beta \leq V(x, X) \leq \beta(|X|^p + 1)$.

Moreover, if $W^{(\delta)}$ fulfil the requirements of Definition (3.1), and if V is as above (with the same α and β), we will call $\{W^{(\delta)}\}_{\delta>0}$ together with V an *admissible family of stored-energy functions for elasticity*.

As presented in Section 1.5, we introduce the family of the corresponding integral functionals

$$\mathcal{E}_\varepsilon^{(\delta)} : L^p(\Omega; \mathbb{R}^n) \to \mathbb{R} \cup \{\infty\}, \quad \delta \geq 0, \quad \varepsilon > 0,$$

defined for $u \in W^{1,p}(\Omega; \mathbb{R}^n)$ as

$$(3.3) \qquad \mathcal{E}_\varepsilon^{(\delta)}(u) := \frac{1}{\delta^p} \int_\Omega W^{(\delta)}\left(\frac{x}{\varepsilon}, I + \delta \nabla u(x)\right) \, dx \quad \text{for } \delta > 0,$$

$$(3.4) \qquad \mathcal{E}_\varepsilon^{(0)}(u) := \int_\Omega V\left(\frac{x}{\varepsilon}, \mathfrak{E}u(x)\right) \, dx,$$

and extended to $L^p(\Omega; \mathbb{R}^n)$ by ∞.

In order for the theory from Chapter 2 to be applicable, we must check the growth and equivalence conditions. Regarding growth, Korn's inequality and its nonlinear counterpart below will provide a link to our theory developed for functionals of Gårding type. Although the following inequality appears to be well-known, it is hard to find a proof in the literature, therefore we include it here.

Theorem (3.5). *Let $\Omega \subset \mathbb{R}^n$ be a bounded Lipschitz domain and $1 < p < \infty$. There exists a constant C, depending only on Ω and p, such that for every $u \in W^{1,p}(\Omega; \mathbb{R}^n)$*

$$\|\nabla u\|_{L^p(\Omega;\mathbb{R}^{n\times n})} \leq C\Big(\|\,\text{dist}\,\big(I + \nabla u, \text{SO}(n)\big)\|_{L^p(\Omega)} + \|u\|_{L^p(\Omega;\mathbb{R}^n)}\Big).$$

Proof. We will show this result by contradiction. Let us suppose that there exists a sequence $\{u_j\}_{j\in\mathbb{N}} \subset W^{1,p}(\Omega; \mathbb{R}^n)$ such that

$$\|\nabla u_j\|_{L^p} > j\Big(\|\,\text{dist}\,\big(I + \nabla u_j, \text{SO}(n)\big)\|_{L^p} + \|u_j\|_{L^p}\Big).$$

Since for every $X \in \mathbb{R}^{n \times n}$ it holds $|X| \leq \text{dist}\left(I + X, \text{SO}(n)\right) + 2\sqrt{n}$, we have for any $u \in W^{1,p}(\Omega; \mathbb{R}^n)$

$$\|\nabla u\|_{L^p} \leq \|\text{dist}\left(I + \nabla u, \text{SO}(n)\right)\|_{L^p} + 2\sqrt{n}|\Omega|^{1/p}.$$

Applying this inequality to our sequence, we obtain

$$2\sqrt{n}|\Omega|^{1/p} > \left(1 - \tfrac{1}{j}\right)\|\nabla u_j\|_{L^p} + \|u_j\|_{L^p}.$$

Hence, $\{u_j\}_{j \in \mathbb{N}}$ is bounded in $W^{1,p}(\Omega; \mathbb{R}^n)$. From our assumption it follows immediately that $u_j \to 0$ in $L^p(\Omega; \mathbb{R}^n)$ and consequently $u_j \rightharpoonup 0$ in $W^{1,p}(\Omega; \mathbb{R}^n)$. Furthermore, by Theorem C.5 there exists a sequence $\{R_j\}_{j \in \mathbb{N}} \subset \text{SO}(n)$ such that

$$\|I + \nabla u_j - R_j\|_{L^p} \leq C_1 \|\text{dist}\left(I + \nabla u_j, \text{SO}(n)\right)\|_{L^p} \leq \tfrac{C_1}{j} \|\nabla u_j\|_{L^p}.$$

Hence $I + \nabla u_j - R_j \to 0$ in $L^p(\Omega; \mathbb{R}^{n \times n})$. This, together with $\nabla u_j \rightharpoonup 0$ in $L^p(\Omega; \mathbb{R}^{n \times n})$, yields $I - R_j \rightharpoonup 0$ in $L^p(\Omega; \mathbb{R}^{n \times n})$. Being a sequence of constant matrices, the last sequence converges even strongly. Therefore, $\nabla u_j \to 0$ in $L^p(\Omega; \mathbb{R}^{n \times n})$, and

$$u_j \to 0 \quad \text{in} \quad W^{1,p}(\Omega; \mathbb{R}^n).$$

Let us, if necessary, pass to a not relabelled subsequence so that $\|\nabla u_j\|_{L^p} \leq \tfrac{1}{j}$. Since $\|I - R_j\|_{L^p} \leq 2C_1\|\nabla u_j\|_{L^p}$, we can, at least from some j_0 on, estimate

$$|I - (R_j)_{\text{sym}}| \leq C_2\|\nabla u_j\|_{L^p}^2 \leq \tfrac{C_2}{j}\|\nabla u_j\|_{L^p}.$$

Hence,

$$\begin{aligned}
\tfrac{C_1}{j}\|\nabla u_j\|_{L^p} &\geq \|I + \nabla u_j - R_j\|_{L^p} \\
&\geq \|I + \mathfrak{E}u_j - (R_j)_{\text{sym}}\|_{L^p} \\
&\geq \|\mathfrak{E}u_j\|_{L^p} - \tfrac{C_2|\Omega|^{1/p}}{j}\|\nabla u_j\|_{L^p}.
\end{aligned}$$

It follows

$$\frac{C_1 + C_2|\Omega|^{1/p} + 1}{j}\|\nabla u_j\|_{L^p} > \|\mathfrak{E}u_j\|_{L^p} + \|u_j\|_{L^p},$$

which contradicts Korn's inequality (see Theorem (C.6)). ∎

Remark (3.6).

(a) Korn's inequality in the second case can be viewed upon as the linearization of the geometric rigidity result in Theorem (C.5) from [FJM:02]. As already announced, when considering growth properties in the linear case, i.e. of V, we will apply Korn's inequality, however, in the general case (see Theorem (C.6)). Therefore, Theorem (3.5) provides the matching "geometric rigidity result in the general case".

(b) In the proof of Lemma (3.7) below, we will use the version with δu being inserted:

$$\|\nabla u\|_{L^p} \leq C\Big(\tfrac{1}{\delta}\|\text{dist}\left(I + \delta\nabla u, \text{SO}(n)\right)\|_{L^p} + \|u\|_{L^p}\Big).$$

Hence, the constant C does not depend on δ.

Now we are in the position to show that an admissible family of stored-energy functions for elasticity generates a family of integral functionals of Gårding type.

Lemma (3.7). *Suppose that*

$$W^{(\delta)} : \mathbb{R}^n \times \mathbb{R}^{n \times n} \to \mathbb{R}, \quad \delta > 0, \quad and \quad V : \mathbb{R}^n \times \mathbb{R}^{n \times n} \to \mathbb{R}$$

make an admissible family of stored-energy functions for elasticity. Then the functions $V^{(\delta)} : \mathbb{R}^n \times \mathbb{R}^{n \times n} \to \mathbb{R}$,

$$V^{(\delta)}(x, X) := \frac{1}{\delta^p} W^{(\delta)}(x, I + \delta X) \quad for \ \delta > 0, \quad and \quad V^{(0)}(x, X) := V(x, X)$$

fulfil the growth conditions in Theorem (2.31).

Proof. For $\delta = 0$, the required upper bound is an assumption. For the lower bound we set $g^{(0)}(X) := \alpha |X_{\text{sym}}|^p - \beta$ and apply Korn's inequality in the general case from Theorem (C.6).

Let $\delta > 0$. It holds

$$V^{(\delta)}(x, X) \le \beta \big(|X|^p + 1 \big)$$

and

$$V^{(\delta)}(x, X) \ge g^{(\delta)}(X) := \frac{\alpha}{\delta^p} \text{dist}^p \big(I + \delta X, \text{SO}(n) \big) - \beta.$$

The required property of $g^{(\delta)}$ follows from Theorem (3.5). ∎

Another requirement for the results of Chapter 2 to hold is equivalence. It should be taken into account that the stored-energy functions possess special symmetry properties, whereas the transition from frame-indifference for $\delta > 0$ to independence from the skew-symmetric part for $\delta = 0$ must be incorporated. With the following lemma we will be able to approximate the values of frame-indifferent functions by the value of the symmetric part of the argument.

Lemma (3.8). *Let* $f : \mathbb{R}^{n \times n} \to \mathbb{R}$ *a frame-indifferent function. There exists* $C > 0$, *dependent only on* n, *such that for every* $X \in \mathbb{R}^{n \times n}$ *there exists* $q(X) \in \mathbb{R}^{n \times n}_{\text{sym}}$ *with*

$$f(I + X) = f(I + X_{\text{sym}} + q(X)) \quad and \quad |q(X)| \le C \min\{|X|, |X|^2\}.$$

The main tool for the proof will be an adapted version of the polar decomposition of a matrix. Let us first explain this.

Any matrix $Y \in \mathbb{R}^{n \times n}$ can be represented as $Y = Q\sqrt{Y^T Y}$ where Q is an orthogonal matrix. If $\det Y > 0$, then even $Q \in \text{SO}(n)$. For Y singular, the matrix Q is not uniquely determined but may be taken from $\text{SO}(n)$ as well, and the same result follows.

Let us consider the remaining case $\det Y < 0$. Here the polar decomposition yields an orthogonal matrix Q with $\det Q = -1$. Let $\sqrt{Y^T Y} = PDP^T$ be some orthogonal diagonalization. If J is the diagonal matrix having -1 as the first diagonal element, all the others being equal to 1, then

$$Y = Q(PDP^T) = (QPJP^T)(PJDP^T).$$

Now $QPJP^T \in \text{SO}(n)$, and $PJDP^T$ is a symmetric matrix.

Hence, an arbitrary $Y \in \mathbb{R}^{n \times n}$ has a decomposition

$$Y = RS$$

with $R \in \mathrm{SO}(n)$ and $S \in \mathbb{R}^{n \times n}_{\mathrm{sym}}$. It always holds $|S| = |Y|$. If $\det Y > 0$, we may take $S = \sqrt{Y^T Y}$.

Proof (of Lemma (3.8)). We will use the decomposition from above for $Y = I + X$. Let S be the corresponding symmetric matrix, and let us write

$$S = I + X_{\mathrm{sym}} + q(X).$$

Obviously

$$f(I + X) = f(I + X_{\mathrm{sym}} + q(X)).$$

For $|X| < 1$ we have $\det(I + X) > 0$, and therefore

$$q(X) = \sqrt{(I + X)^T (I + X)} - I - X_{\mathrm{sym}}.$$

There exists a constant $C > 0$ such that

$$|q(X)| \leq C \min\{|X|, |X|^2\}.$$

This inequality holds (possibly by adjusting C) also for $|X| \geq 1$, as in that case

$$|q(X)| = |S - I - X_{\mathrm{sym}}| \leq |S| + |I + X| = 2|I + X| \leq 2(\sqrt{n} + 1)|X|. \quad \blacksquare$$

3.2 Homogenization and geometric linearization

Our main result is the following theorem on homogenization and geometric linearization for multiwell energy functionals. In particular we will see that these limiting processes commute and may be taken simultaneously.

Theorem (3.9). *Let*

$$W^{(\delta)} : \mathbb{R}^n \times \mathbb{R}^{n \times n} \to \mathbb{R}, \quad \delta > 0, \quad with \quad V : \mathbb{R}^n \times \mathbb{R}^{n \times n} \to \mathbb{R}$$

be an admissible family of stored-energy functions for elasticity. Assume that

- *every $W^{(\delta)}$ is homogenizable,*

- *V is Carathéodory,*

- *for all $R > 0$*

$$(3.10) \quad \limsup_{\delta \to 0} \sup_{T > 0} \frac{1}{(2T)^n} \int_{(-T,T)^n} \sup_{\substack{X \in \mathbb{R}^{n \times n}_{\mathrm{sym}} \\ |X| \leq R}} \left| \tfrac{1}{\delta^p} W^{(\delta)}(x, I + \delta X) - V(x, X) \right| \, dx = 0$$

and

$$(3.11) \quad \lim_{|Z| \to 0} \sup_{T > 0} \frac{1}{(2T)^n} \int_{(-T,T)^n} \sup_{|X| \leq R} |V(x, X + Z) - V(x, X)| \, dx = 0.$$

Then also V is homogenizable. The limiting density V_{hom} is given by

$$V_{\mathrm{hom}}(X) = \lim_{k\to\infty} \inf \left\{ \frac{1}{k^n} \int_{k\mathbb{I}^n} V(x, X + \mathfrak{E}\varphi(x))\, dx : \varphi \in W_0^{1,p}(k\mathbb{I}^n; \mathbb{R}^n) \right\}$$

and satisfies

$$V_{\mathrm{hom}}(X) = \lim_{\delta\to 0} \frac{1}{\delta^p} W_{\mathrm{hom}}^{(\delta)}(I + \delta X)$$

for every $X \in \mathbb{R}^{n\times n}$. Moreover, the following diagram of the corresponding integral functionals commutes for every $U \in \mathcal{A}(\mathbb{R}^n)$:

$$
\begin{array}{ccc}
\mathcal{E}_\varepsilon^{(\delta)}(\_,U) & \xdashrightarrow{\ \Gamma\ } & \mathrm{lsc}\,\mathcal{E}_\varepsilon^{(0)}(\_,U) \\[4pt]
\Big\downarrow{\scriptstyle \Gamma} & & \Big\downarrow{\scriptstyle \Gamma} \\[4pt]
\mathcal{E}_{\mathrm{hom}}^{(\delta)}(\_,U) & \xdashrightarrow{\ \Gamma\ } & \mathcal{E}_{\mathrm{hom}}^{(0)}(\_,U)
\end{array}
$$

Here $\mathcal{E}_\varepsilon^{(\delta)}$, $\delta \geq 0$, are as in (3.3) and (3.4), and

$$\mathcal{E}_{\mathrm{hom}}^{(0)}(u,U) = \int_U V_{\mathrm{hom}}(\mathfrak{E}u(x))\, dx,$$

$$\mathrm{lsc}\,\mathcal{E}_\varepsilon^{(0)}(u,U) = \int_U V^{\mathrm{qc}}\!\left(\frac{x}{\varepsilon}, \mathfrak{E}u(x)\right) dx$$

for $u \in W^{1,p}(\Omega; \mathbb{R}^n)$ and ∞ otherwise. In fact, for any $\varepsilon_k \searrow 0$, $\delta_k \searrow 0$

$$\Gamma(L^p)\text{-}\lim_{k\to\infty} \mathcal{E}_{\varepsilon_k}^{(\delta_k)}(\_,U) = \mathcal{E}_{\mathrm{hom}}^{(0)}(\_,U).$$

Remark (3.12).

(a) Comparing the assumptions with those of Theorem (1.1), it follows from Corollary (2.33) that we replaced periodicity with a more general property: homogenizability. The equivalence condition (3.10) is for $\mathbb{I}^n$-periodic functions weaker than in Theorem (1.1) and is again imposed only on symmetric matrices. The extension to all matrices will follow from the symmetry properties. Lastly, the condition (3.11) is a mild continuity assumption and is fulfilled for $\mathbb{I}^n$-periodic Carathéodory functions. Namely, by periodicity the condition has to be proved for a fixed T, say $T = 1$. This may be done as in the proof of Theorem (3.15).

(b) We showed in Corollary (2.24) that $\mathrm{lsc}\,\mathcal{E}_\varepsilon^{(0)}$ has indeed density V^{qc} (of course with ε in the denominator of the first argument). According to the relation between quasiconvex envelopes in Appendix A,

$$V^{\mathrm{qc}}(\tfrac{x}{\varepsilon}, X) = V^{\mathrm{qc}}(\tfrac{x}{\varepsilon}, X_{\mathrm{sym}}) = \left(V|_{\mathbb{R}^n \times \mathbb{R}^{n\times n}_{\mathrm{sym}}}\right)^{\mathrm{qcls}}(\tfrac{x}{\varepsilon}, X_{\mathrm{sym}}).$$

Therefore, the functional $\mathrm{lsc}\,\mathcal{E}_\varepsilon^{(0)}$ coincides with $\mathcal{E}_\varepsilon^{(\mathrm{rel})}$ in Theorem (1.1).

(c) If $W_\delta = W$ admits a quadratic Taylor expansion, we get a result closely related to those in [MN:11, GN:11].

(d) A straightforward adaptation of Theorem (2.19) yields a version of Theorem (3.9) for the functionals $\bar{\mathcal{E}}_\varepsilon^{(\delta)}$ with prescribed boundary values.

Proof. Define $V^{(\delta)}$, $\delta \geq 0$, as in Lemma (3.7). We have to check that these densities meet the conditions of Theorems (2.31) and (2.34). According to Lemma (3.7), the growth conditions are satisfied. Also, if $W^{(\delta)}$ is homogenizable, $V^{(\delta)}$ is homogenizable as well.

As for the equivalence, we assumed it only on the symmetric matrices, i.e., for any $R > 0$

$$\limsup_{\delta \to 0} \sup_{T>0} \frac{1}{(2T)^n} \int_{(-T,T)^n} \sup_{\substack{X \in \mathbb{R}^{n \times n}_{\mathrm{sym}} \\ |X| \leq R}} \left| V^{(\delta)}(x,X) - V(x,X) \right| \, dx = 0.$$

For an arbitrary $X \in \mathbb{R}^{n \times n}$ we may, according to Lemma (3.8), write

$$V^{(\delta)}(x,X) = V^{(\delta)}(x, X_{\mathrm{sym}} + \tfrac{1}{\delta}q(\delta X)).$$

Therefore,

$$\begin{aligned} V^{(\delta)}(x,X) - V(x,X) &= V^{(\delta)}(x, X_{\mathrm{sym}} + \tfrac{1}{\delta}q(\delta X)) - V(x, X_{\mathrm{sym}} + \tfrac{1}{\delta}q(\delta X)) + \\ &\quad + V(x, X_{\mathrm{sym}} + \tfrac{1}{\delta}q(\delta X)) - V(x, X_{\mathrm{sym}}). \end{aligned}$$

Fix $R > 0$. If $|X| \leq R$, then $\frac{1}{\delta}|q(\delta X)| \leq C|X| \leq CR$. Hence,

$$\sup_{|X| \leq R} \left| V^{(\delta)}(x, X_{\mathrm{sym}} + \tfrac{1}{\delta}q(\delta X)) - V(x, X_{\mathrm{sym}} + \tfrac{1}{\delta}q(\delta X)) \right| \leq \sup_{\substack{Y \in \mathbb{R}^{n \times n}_{\mathrm{sym}} \\ |Y| \leq (C+1)R}} \left| V^{(\delta)}(x,Y) - V(x,Y) \right|.$$

To assess $\left| V(x, X_{\mathrm{sym}} + \tfrac{1}{\delta}q(\delta X)) - V(x, X_{\mathrm{sym}}) \right|$, we use the other bound

$$\tfrac{1}{\delta}q(\delta X) \leq C\delta|X|^2 \leq C\delta R^2.$$

Thus, it follows from (3.11)

$$\begin{aligned} &\limsup_{\delta \to 0} \sup_{T>0} \frac{1}{(2T)^n} \int_{(-T,T)^n} \sup_{|X| \leq R} \left| V(x, X_{\mathrm{sym}} + \tfrac{1}{\delta}q(\delta X)) - V(x, X_{\mathrm{sym}}) \right| \, dx \\ &\leq \lim_{|Z| \to 0} \sup_{T>0} \frac{1}{(2T)^n} \int_{(-T,T)^n} \sup_{|X| \leq R} \left| V(x, X_{\mathrm{sym}} + Z) - V(x, X_{\mathrm{sym}}) \right| \, dx \\ &= 0. \end{aligned}$$

Hence, the equivalence condition

$$\limsup_{\delta \to 0} \sup_{T>0} \frac{1}{(2T)^n} \int_{(-T,T)^n} \sup_{|X| \leq R} \left| V^{(\delta)}(x,X) - V(x,X) \right| \, dx = 0$$

has been proved, and we may apply Theorem (2.34). Moreover, even the assumptions of Theorem (2.26) are met, and the result on simultaneous limits follows. ∎

3.3   Additional comments

Our definition of an admissible family of stored-energy functions for elasticity is compatible with homogenization in the following sense:

Proposition (3.13). *Suppose homogenizable functions $W^{(\delta)}$, $\delta > 0$, and V make an admissible family of stored-energy functions for elasticity. Then $W^{(\delta)}_{\mathrm{hom}}$, $\delta > 0$, with V_{hom} is again such family.*

The only nontrivial part of the proof will be the lower bound for $\delta > 0$. Therefore, we extend the Zhang's result for $p = 2$ (see Theorem 1.1 in [Zh:97]) to general $p > 1$.

Theorem (3.14). *For $1 < p < \infty$ it holds*

$$[\mathrm{dist}^p(\_,\mathrm{SO}(n))]^{\mathrm{qc}} \geq \frac{1}{C^p_{\mathrm{geo}}} \,\mathrm{dist}^p(\_,\mathrm{SO}(n))$$

where C_{geo} is the constant from the geometric rigidity result for unit cube $\mathbb{I}^n$ and for the exponent p.

Proof. By Theorem (C.5), for any $\varphi \in C^\infty_c(\mathbb{I}^n)$ there exists an $R \in \mathrm{SO}(n)$ such that

$$
\begin{aligned}
C^p_{\mathrm{geo}} \int_{\mathbb{I}^n} \mathrm{dist}^p(X + \nabla\varphi(x), \mathrm{SO}(n)) \, dx
&\geq \int_{\mathbb{I}^n} |X + \nabla\varphi(x) - R|^p \, dx \\
&\geq \left| \int_{\mathbb{I}^n} (X + \nabla\varphi(x) - R) \, dx \right|^p \\
&= |X - R|^p \\
&\geq \mathrm{dist}^p(X, \mathrm{SO}(n)),
\end{aligned}
$$

where in the second step we have applied Jensen's inequality. Consequently,

$$
\begin{aligned}
\left[\mathrm{dist}^p(\_,\mathrm{SO}(n))\right]^{\mathrm{qc}}(X)
&= \inf_{\varphi \in C^\infty_c(\mathbb{I}^n)} \int_{\mathbb{I}^n} \mathrm{dist}^p(X + \nabla\varphi(x), \mathrm{SO}(n)) \, dx \\
&\geq \frac{1}{C^p_{\mathrm{geo}}} \,\mathrm{dist}^p(X, \mathrm{SO}(n)). \quad\blacksquare
\end{aligned}
$$

Proof (of Proposition (3.13)). Clearly, for every $\delta > 0$ the function $W^{(\delta)}_{\mathrm{hom}}$ is frame-indifferent, and V_{hom} depends only on the symmetric part of the argument. The upper bounds are also easy to see. The lower bound for $W^{(\delta)}_{\mathrm{hom}}$ follows from Theorem (3.14) as

$$
\begin{aligned}
W^{(\delta)}_{\mathrm{hom}}
&\geq [\alpha \,\mathrm{dist}^p(\_,\mathrm{SO}(n)) - \beta\delta^p]_{\mathrm{hom}} \\
&= \alpha \,[\mathrm{dist}^p(\_,\mathrm{SO}(n))]^{\mathrm{qc}} - \beta\delta^p \\
&\geq \frac{\alpha}{C^p_{\mathrm{geo}}} \,\mathrm{dist}^p(\_,\mathrm{SO}(n)) - \beta\delta^p.
\end{aligned}
$$

For V we use Jensen's inequality that yields

$$\frac{1}{k^n} \int_{k\mathbb{I}^n} V(x, X + \nabla\varphi(x)) \, dx \geq \frac{1}{k^n} \int_{k\mathbb{I}^n} (\alpha|X + \nabla\varphi(x)|^p - \beta) \, dx \geq \alpha|X|^p - \beta$$

for all $k \in \mathbb{N}$ and $\varphi \in W^{1,p}_0(k\mathbb{I}^n; \mathbb{R}^n)$. $\blacksquare$

Let us at the end mention that the geometric linearization result of [Sch:08] for the multiwell case itself follows from our theory developed in Chapter 2 (and consequently, so does the basic result from [DNP:02] for the one-well case). Since this fact appears only implicitly in Theorem (3.9), see the upper row of the diagram, and since no periodicity or homogenizability is needed, let us state and prove this claim clearly. However, compared to [DNP:02] and [Sch:08], we assume an upper bound on $V^{(\delta)}$. This could be circumvented, see also Remark (2.11).

Proposition (3.15). *Let $\Omega \subset \mathbb{R}^n$ be a bounded Lipschitz domain. Let us have a family of frame-indifferent Borel functions*

$$W^{(\delta)} : \Omega \times \mathbb{R}^{n \times n} \to \mathbb{R}, \quad \delta > 0,$$

and denote

$$V^{(\delta)} : \Omega \times \mathbb{R}^{n \times n}_{\mathrm{sym}} \to \mathbb{R}, \quad V^{(\delta)}(x, Y) := \frac{1}{\delta^p} W^{(\delta)}(x, I + \delta Y).$$

Suppose there exist $\alpha, \beta > 0$ such that for every $\delta > 0$ and for a.e. $x \in \Omega$

- $W^{(\delta)}(x, X) \geq \alpha \operatorname{dist}^p(X, \mathrm{SO}(n)) - \beta \, \delta^p$ *for all $X \in \mathbb{R}^{n \times n}$.*

- $V^{(\delta)}(x, X) \leq \beta(|X|^p + 1)$ *for all $X \in \mathbb{R}^{n \times n}_{\mathrm{sym}}$,*

and assume that the family $V^{(\delta)}$, $\delta > 0$, converges as $\delta \to 0$ to a Carathéodory function

$$V : \Omega \times \mathbb{R}^{n \times n}_{\mathrm{sym}} \to \mathbb{R}$$

uniformly on $\Omega \times K$ for every compact $K \subset \mathbb{R}^{n \times n}_{\mathrm{sym}}$. Then

$$\Gamma\text{-}\lim_{\delta \to 0} \mathcal{E}^{(\delta)} = \mathcal{E}^{(\mathrm{rel})}$$

where for $\delta > 0$

$$\mathcal{E}^{(\delta)}(u) := \begin{cases} \frac{1}{\delta^p} \int_\Omega W^{(\delta)}(x, I + \delta \nabla u(x)) \, dx, & u \in W^{1,p}(\Omega; \mathbb{R}^n), \\ \infty, & else, \end{cases}$$

and

$$\mathcal{E}^{(\mathrm{rel})}(u) := \begin{cases} \int_\Omega V^{\mathrm{qcls}}(x, \mathfrak{E}u(x)) \, dx, & u \in W^{1,p}(\Omega; \mathbb{R}^n), \\ \infty, & else. \end{cases}$$

Proof. We will apply the relaxation result from Theorem (2.23). The growth conditions for

$$f^{(\delta)}(x, X) := \frac{1}{\delta^p} W^{(\delta)}(x, I + \delta X), \quad f(x, X) := V(x, X_{\mathrm{sym}})$$

follow mostly by the same arguing as in Lemma (3.7). Additionally, for the lower bound for f, we apply

$$\lim_{\delta \to 0} \frac{1}{\delta^p} \operatorname{dist}^p(I + \delta X, \mathrm{SO}(n)) = |X_{\mathrm{sym}}|^p$$

whereas the upper bound for $f^{(\delta)}$ is a consequence of Lemma (3.8) as

$$f^{(\delta)}(x, X) = V^{(\delta)}(x, X_{\mathrm{sym}} + \tfrac{1}{\delta} q(\delta X)) \leq \beta\big(1 + |X_{\mathrm{sym}} + \tfrac{1}{\delta} q(\delta X)|^p\big) \leq C\big(1 + |X|^p\big).$$

We still have to prove the equivalence condition, i.e., that for any $R > 0$

$$\lim_{\delta \to 0} \int_{\Omega} \sup_{|X| \le R} \left| f^{(\delta)}(x, X) - f(x, X) \right| \, dx = 0.$$

Analogous as in the proof of Theorem (3.9), we write

$$f^{(\delta)}(x, X) - f(x, X) = V^{(\delta)}(x, X_{\text{sym}} + \tfrac{1}{\delta} q(\delta X)) - V(x, X_{\text{sym}} + \tfrac{1}{\delta} q(\delta X)) +$$
$$+ V(x, X_{\text{sym}} + \tfrac{1}{\delta} q(\delta X)) - V(x, X_{\text{sym}}).$$

For the first difference,

$$\lim_{\delta \to 0} \int_{\Omega} \sup_{|X| \le R} \left| V^{(\delta)}(x, X_{\text{sym}} + \tfrac{1}{\delta} q(\delta X)) - V(x, X_{\text{sym}} + \tfrac{1}{\delta} q(\delta X)) \right| \, dx = 0$$

follows from the local uniform convergence.

Fix $R > 0$. Again,

$$|X| \le R \Rightarrow \tfrac{1}{\delta} q(\delta X) \le C \delta R^2.$$

Choose arbitrary $\eta > 0$. Since V is Carathéodory, there exists by Scorza-Dragoni's theorem (see, e.g., Theorem 6.35 in [FL:07]) a compact set $K \subset \Omega$ such that $|\Omega \setminus K| < \eta$ and $V|_{K \times \mathbb{R}^{n \times n}_{\text{sym}}}$ is continuous. Hence, V is uniformly continuous on

$$K \times \{ X \in \mathbb{R}^{n \times n}_{\text{sym}} : |X| \le R + CR^2 \}.$$

There exists $\delta_\eta < 1$ such that for all $x \in K$ and $\delta < \delta_\eta$

$$\sup_{|X| \le R} \left| V(x, X_{\text{sym}} + \tfrac{1}{\delta} q(\delta X)) - V(x, X_{\text{sym}}) \right| < \eta.$$

Moreover,

$$\int_{\Omega \setminus K} \sup_{|X| \le R} \left| V(x, X_{\text{sym}} + \tfrac{1}{\delta} q(\delta X)) - V(x, X_{\text{sym}}) \right| \, dx < 2\beta(1 + (C+1)^p R^p)\eta.$$

Thus,

$$\lim_{\delta \to 0} \int_{\Omega} \sup_{|X| \le R} \left| V(x, X_{\text{sym}} + \tfrac{1}{\delta} q(\delta X)) - V(x, X_{\text{sym}}) \right| \, dx = 0.$$

Hence, the equivalence condition has been proved. Since now all conditions of Theorem (2.23) are met,

$$\Gamma(L^p)\text{-} \lim_{\delta \to 0} \mathcal{E}^{(\delta)} = \operatorname{lsc} \mathcal{E}$$

where

$$\mathcal{E}(u) = \begin{cases} \int_{\Omega} f(x, \nabla u(x)) \, dx, & u \in W^{1,p}(\Omega; \mathbb{R}^n), \\ \infty, & \text{else.} \end{cases}$$

According to Corollary (2.24), $\operatorname{lsc} \mathcal{E}$ has density f^{qc}. By Zhang's result (see Appendix A),

$$f^{\text{qc}}(x, X) = V^{\text{qcls}}(x, X_{\text{sym}}). \quad \blacksquare$$

Chapter 4

Stochastic homogenization

In Section 2.7 we extended the validity of the known results in homogenization. That leads us to the idea of exploring the corresponding stochastic setting. Namely, there are related results for random integral functionals, starting with a couple of articles [DM:86-1, DM:86-2] from Dal Maso and Modica, where they consider convex densities. For non-convex densities, this was done by Messaoudi and Michaille in [MM:94].

In this chapter we will first present the fundamentals of the ergodic theory with the ergodic theorem, originally proved in [AK:81]. Beside the mentioned articles, we will also use the survey paper [LM:02] as a source (see also Section 12.4 in [ABM:06]). After introducing necessary terminology and denotations, we will turn our attention to homogenization. In Theorem (4.9) we will state the essential result from [MM:94] that corresponds to the periodic homogenization in the deterministic case.

Our own results start with extending Theorem (4.9) to densities that do not fulfil the local Lipschitz condition (4.10) by passing to the lower semicontinuous envelope and using Yosida's transform. Then we will, as in Chapter 2, also allow for lower bounds of Gårding type. We will conclude the chapter with stochastic versions of the homogenization closure and the commutability result. We will replace the assumptions of periodicity in law and ergodicity with homogenizability, as we did with $\mathbb{I}^n$-periodicity in the deterministic case.

4.1   Ergodic theory

Let us have a probability space $(\Xi, \mathcal{S}, P)$, i.e., a set Ξ with a σ-algebra $\mathcal{S}$ and a probability measure P. We call $\tau = \{\tau_z : \Xi \to \Xi\}_{z \in \mathbb{Z}^n}$ a *measurable group of transformations on* Ξ if

- τ_z is measurable for every $z \in \mathbb{Z}^n$,
- $(\tau, \circ)$ is a group with $\tau_z \circ \tau_w = \tau_{z+w}$ and $\tau_z^{-1} = \tau_{-z}$ for all $z, w \in \mathbb{Z}^n$.

We say that these transformations are *measure-preserving* if $P(\tau_z(E)) = P(E)$ for every $z \in \mathbb{Z}^n$ and $E \in \mathcal{S}$. The group τ is said to be *ergodic* if in addition for each $E \in \mathcal{S}$ it holds

$$\tau_z(E) = E \quad \text{for every } z \in \mathbb{Z}^n \quad \Longrightarrow \quad P(E) \in \{0, 1\}.$$

Let us denote by $\mathcal{B}_b(\mathbb{R}^n)$ the set of all bounded Borel sets in $\mathbb{R}^n$. $\mathcal{Z}^n$ will stand for its subset that consists of all rectangular cuboids $[a_1, b_1) \times \ldots \times [a_n, b_n)$ with $a_i, b_i \in \mathbb{Z}$ for all $i = 1, \ldots, n$. For the unit interval of this type, let us write $\mathbb{I}_0 := [0, 1) = \mathbb{I} \cup \{0\}$.

Definition (4.1). Take a probability space $(\Xi, \mathcal{S}, P)$ and a measurable group of measure preserving transformations $\tau = \{\tau_z\}_{z \in \mathbb{Z}^n}$. A set function

$$\Upsilon : \mathcal{B}_b(\mathbb{R}^n) \to L^1(\Xi)$$

is a *subadditive process with respect to τ* if it is

- *subadditive*, i.e., for arbitrary disjoint $A, B \in \mathcal{B}_b(\mathbb{R}^n)$ it holds

$$\Upsilon(A \cup B) \leq \Upsilon(A) + \Upsilon(B),$$

- *covariant*, i.e., for all $z \in \mathbb{Z}^n$ and $A \in \mathcal{B}_b(\mathbb{R}^n)$

$$\Upsilon(A) \circ \tau_z = \Upsilon(z + A),$$

- the *spatial constant*

$$\gamma(\Upsilon) := \inf_{I \in \mathcal{Z}^n, \, |I| > 0} \frac{1}{|I|} \int_\Xi \Upsilon(I)(\xi) \, dP(\xi)$$

  is finite.

For such a process the spatial constant can be (e.g., by (3.4) Lemma in [AK:81]) calculated as the limit

$$\gamma(\Upsilon) = \lim_{k \to \infty} \frac{1}{k^n} \int_\Xi \Upsilon(k\mathbb{I}_0^n)(\xi) \, dP(\xi).$$

If we have such a process, then we immediately can say something about its convergence properties. Let us state Theorem 4.1 of [LM:02].

Theorem (4.2). *Let $\Upsilon : \mathcal{B}_b(\mathbb{R}^n) \to L^1(\Xi)$ be a subadditive process with respect to τ, and suppose that there exists $b \in L^1(\Xi)$ such that $|\Upsilon(A)| \leq b$ for every Borel set $A \subset \mathbb{I}_0^n$. For every regular sequence of convex Borel sets $\{A_k\}_{k \in \mathbb{N}}$ in $\mathbb{R}^n$ with $\lim_{k \to \infty} \rho(A_k) = \infty$, it holds:*

(a) *The sequence $\{\frac{1}{|A_k|} \Upsilon(A_k)\}_{k \in \mathbb{N}}$ converges almost surely.*

(b) *If τ is even ergodic, then the limit is the constant function $\gamma(\Upsilon)$. More precisely, there exists $\Xi' \subset \Xi$ with $P(\Xi') = 1$ such that*

$$\lim_{k \to \infty} \frac{1}{|A_k|} \Upsilon(A_k)(\xi) = \gamma(\Upsilon).$$

for every $\xi \in \Xi'$.

Remark (4.3).

(a) In our applications we will mostly have sequences of the form $\{l_k A\}_{k \in \mathbb{N}}$ with a single Borel convex set A with nonempty interior and $l_k \nearrow \infty$. They fulfil the assumptions in the theorem, but let us nonetheless explain the terminology. A sequence of sets $\{A_k\}_{k \in \mathbb{N}}$ is *regular* if there exist an increasing sequence $\{I_k\}_{k \in \mathbb{N}} \subset \mathcal{Z}^n$ and $C > 0$ (independent of k) such that

$$A_k \subset I_k \quad \text{and} \quad |I_k| \leq C|A_k|$$

for all $k \in \mathbb{N}$. Furthermore, $\rho(A)$ is the radius of the largest closed ball that lies within A.

(b) This result from [LM:02] generalizes the first ergodic theorem of this type from [AK:81], where only regular sequences of cuboids from $\mathcal{Z}^n$ were considered.

In the stochastic setting, integral functionals depend also on the stochastic variable ξ. In other words, we will have maps from probability spaces to sets consisting of integral functionals. For further analysis, the latter should be made measurable spaces since the dependence on ξ must be measurable.

Let us denote by $\mathscr{F}^p_{\alpha,\beta}$ the set of all integral functionals fulfilling the standard p-growth (for now any $p \geq 1$) with the coefficients $\beta \geq \alpha > 0$. More precisely, let us have an integral functional

$$\mathcal{F} : L^p(\mathbb{R}^n; \mathbb{R}^m) \times \mathcal{O}(\mathbb{R}^n) \to \mathbb{R},$$

where $\mathcal{O}(\mathbb{R}^n)$ stands for the family of all open bounded subsets of $\mathbb{R}^n$. Then $\mathcal{F} \in \mathscr{F}^p_{\alpha,\beta}$ if and only if there exists a Carathéodory function $f : \mathbb{R}^n \times \mathbb{R}^{m \times n} \to \mathbb{R}$ with

$$\alpha |X|^p \leq f(x, X) \leq \beta(|X|^p + 1)$$

for a.e. $x \in \mathbb{R}^m$ and all $X \in \mathbb{R}^{m \times n}$ such that

$$\mathcal{F}(u, U) = \begin{cases} \int_U f(x, \nabla u(x)) \ dx, & u \in W^{1,p}(\mathbb{R}^n; \mathbb{R}^m), \\ \infty, & \text{else}, \end{cases}$$

for every $u \in L^p(\mathbb{R}^m; \mathbb{R}^n)$ and $U \in \mathcal{O}(\mathbb{R}^m)$. For the sake of simplicity, we decided to consider only non-negative lower bounds.

Moreover, let $\mathscr{F}^p_\beta$ consist of all integral functionals that correspond to functions with

$$0 \leq f(x, X) \leq \beta(|X|^p + 1)$$

for a.e. $x \in \mathbb{R}^m$ and all $X \in \mathbb{R}^{m \times n}$.

Finally, let

$$\mathscr{F}^p := \bigcup_{\beta \in (0,\infty)} \mathscr{F}^p_\beta.$$

The set $\mathscr{F}^p$ can be viewed upon as a subset of the product space $\mathbb{R}^{W^{1,p}(\mathbb{R}^n; \mathbb{R}^m) \times \mathcal{O}(\mathbb{R}^n)}$. On this space there exists the product σ-algebra. This is the smallest σ-algebra such that all projections are measurable, i.e., the evaluation

$$\mathbb{R}^{W^{1,p}(\mathbb{R}^n; \mathbb{R}^m) \times \mathcal{O}(\mathbb{R}^n)} \to \mathbb{R}, \quad \mathcal{F} \mapsto \mathcal{F}(u, U),$$

is measurable for each $u \in W^{1,p}(\mathbb{R}^n; \mathbb{R}^m)$ and $U \in \mathcal{O}(\mathbb{R}^n)$. (We are taking the Borel σ-algebra on $\mathbb{R}$.) Let $\mathscr{F}^p$ be henceforth equipped with the induced σ-algebra $\mathcal{P}(\mathscr{F}^p)$.

To consider an analogue of the periodicity in the stochastic case, let us define for each $z \in \mathbb{Z}^n$

$$\tau_z : \mathscr{F}^p \to \mathscr{F}^p, \quad (\tau_z \mathcal{F})(u, U) := \int_U f(x + z, \nabla u(x)) \ dx.$$

It is elementary to check that $\tau := \{\tau_z\}_{z \in \mathbb{Z}^n}$ is a measurable group of transformations and

$$\tau_z(\mathscr{F}^p_{\alpha,\beta}) = \mathscr{F}^p_{\alpha,\beta} \quad \text{and} \quad \tau_z(\mathscr{F}^p_\beta) = \mathscr{F}^p_\beta$$

for every $z \in \mathbb{Z}^n$ and $\alpha, \beta > 0$.

Definition (4.4). A *random integral functional* is a measurable mapping

$$\Phi : \Xi \to \mathscr{F}^p$$

such that for some $\beta > 0$ it holds $\Phi(\xi) \subset \mathscr{F}^p_\beta$ for almost every $\xi \in \Xi$.

Henceforth we will often use the sloppy denotation $\Phi : \Xi \to \mathscr{F}^p_\beta$ or $\Phi : \Xi \to \mathscr{F}^p_{\alpha,\beta}$.

If Φ is a random integral functional, then the pointwisely defined mapping $\tau_z \Phi$ is also a random integral functional for any $z \in \mathbb{Z}^n$. We say that the random integral is *periodic in law* if for any $E \in \mathcal{P}(\mathscr{F}^p)$ and $z \in \mathbb{Z}^n$

$$P(\tau_z \Phi \in E) = P\big(\{\xi \in \Xi : \tau_z \Phi(\xi) \in E\}\big) = P\big(\{\xi \in \Xi : \Phi(\xi) \in E\}\big) = P(\Phi \in E)$$

or, in terms of push-forward measures,

$$(\tau_z \Phi)_* P = \Phi_* P.$$

If we equip $\big(\mathscr{F}^p, \mathcal{P}(\mathscr{F}^p)\big)$ with the measure $\Phi_* P$, then

$$\Phi \text{ periodic in law } \iff \tau_z \text{ is measure-preserving}$$

since

$$(\tau_z \Phi)_* P = \Phi_* P \circ \tau_{-z}.$$

A random integral functional Φ is said to be *ergodic* if τ is ergodic for $\Phi_* P$, i.e., for every $E \in \mathcal{P}(\mathscr{F}^p)$ it holds

$$\tau_z(E) = E \quad \text{for every } z \in \mathbb{Z}^n \quad \implies \quad \Phi_* P(E) = P\big(\{\xi \in \Xi : \Phi(\xi) \in E\}\big) \in \{0,1\}.$$

Let us for all $X \in \mathbb{R}^{m \times n}$, $\mathcal{F} \in \mathscr{F}^p$ and $U \in \mathcal{O}(\mathbb{R}^n)$ define

$$M(X; \mathcal{F}, U) := \inf_{\varphi \in W_0^{1,p}(U;\mathbb{R}^m)} \mathcal{F}(\ell_X + \varphi, U),$$

where $\ell_X(x) = X\,x$, and let us denote the same value in terms of densities by

$$m(X; f, U) := \inf_{\varphi \in W_0^{1,p}(U;\mathbb{R}^m)} \int_U f(x, X + \nabla\varphi(x))\,dx.$$

(In the trivial case $U = \emptyset$, we set the value 0. We regard $W_0^{1,p}(U;\mathbb{R}^m)$ as a subspace of $W^{1,p}(\mathbb{R}^n;\mathbb{R}^m)$ through the extension by the zero function.) To consider the measurability of the defined maps, we will need the following working lemma.

Lemma (4.5). *Suppose* $\mathcal{F} \in \mathscr{F}^p$ *and* $U \in \mathcal{O}(\mathbb{R}^n)$. *Let* $\mathcal{Y} \subset W^{1,p}(\mathbb{R}^n;\mathbb{R}^m)$ *be a closed subspace and* $\mathcal{D} \subset \mathcal{Y}$ *a dense subset. Then for every* $u \in W^{1,p}(\mathbb{R}^n;\mathbb{R}^m)$ *it holds*

$$\inf_{\varphi \in \mathcal{Y}} \mathcal{F}(u + \varphi, U) = \inf_{\varphi \in \mathcal{D}} \mathcal{F}(u + \varphi, U).$$

Therefore, if $\mathcal{D}$ *is dense in* $W_0^{1,p}(U;\mathbb{R}^m)$, *then*

$$M(X; \mathcal{F}, U) = \inf_{\varphi \in \mathcal{D}} \mathcal{F}\big(\ell_X + \varphi, U\big)$$

for every $X \in \mathbb{R}^{m \times n}$.

Proof. Take any $\varphi \in \mathcal{Y}$. We may find a sequence $\{\varphi_j\}_{j\in\mathbb{N}} \subset \mathcal{D}$ such that

$$\varphi_j \to \varphi \quad \text{in } W^{1,p}(U;\mathbb{R}^m) \quad \text{and} \quad \lim_{j\to\infty} \nabla\varphi_j(x) = \nabla\varphi(x) \quad \text{for a.e. } x \in U.$$

We may further demand that there exists $v \in L^p(U)$ such that $|\nabla\varphi_j| \leq v$ for all $j \in \mathbb{N}$ (see, e.g., Theorem 2.20 in [FL:07]). Hence, with an appropriate β,

$$f(x, \nabla u(x) + \nabla\varphi_j(x)) \leq \beta(|\nabla u(x) + \nabla\varphi_j(x)|^p + 1) \leq 2^{p-1}\beta(1 + |\nabla u(x)|^p + v(x)^p).$$

Therefore, we may apply the dominated convergence theorem, which yields

$$\lim_{j\to\infty} \int_U f(x, \nabla u(x) + \nabla\varphi_j(x))\, dx = \int_U f(x, \nabla u(x) + \nabla\varphi(x))\, dx. \quad \blacksquare$$

By the definition, the evaluation $\mathcal{F} \mapsto \mathcal{F}(u, U)$ is measurable for every $u \in W^{1,p}(\mathbb{R}^n; \mathbb{R}^m)$ and $U \in \mathcal{O}(\mathbb{R}^n)$. Since $W_0^{1,p}(U;\mathbb{R}^m)$ is separable, the lemma above implies that

$$M(X; \_, U) : \mathscr{F}^p \to \mathbb{R}$$

is a measurable function for every $X \in \mathbb{R}^{m\times n}$ and $U \in \mathcal{O}(\mathbb{R}^n)$. We notice also

$$0 \leq M(X; \_, U)|_{\mathscr{F}^p_\beta} \leq \beta|U|(|X|^p + 1).$$

Proposition (4.6). *Let us have a random integral functional $\Phi : \Xi \to \mathscr{F}^p_\beta$ that is periodic in law. For every $X \in \mathbb{R}^{m\times n}$ the map*

$$\Upsilon_X : \mathcal{B}_b(\mathbb{R}^n) \to L^1(\mathscr{F}^p_\beta, \mathcal{P}(\mathscr{F}^p_\beta), \Phi_*P), \quad \Upsilon_X(A) := M(X; \_, \text{Int } A)$$

is a subadditive process with respect to τ on $\mathscr{F}^p_\beta$. Moreover, the following equiboundedness property holds:

$$\|\Upsilon_X(A)\|_{L^1(\mathscr{F}^p_\beta)} \leq \beta|A|(|X|^p + 1).$$

Proof. First, the map is well-defined as we have already proved that for any $A \in \mathcal{B}_b(\mathbb{R}^n)$ indeed $\Upsilon_X(A) \in L^\infty(\mathscr{F}^p_\beta) \subset L^1(\mathscr{F}^p_\beta)$. Now we prove the three conditions:

- Let us have $A_1, A_2 \in \mathcal{B}_b(\mathbb{R}^n)$ with $A_1 \cap A_2 = \emptyset$. Denote $A := A_1 \cup A_2$. For every choice of $\varphi_i \in W_0^{1,p}(\text{Int } A_i)$, we may view their sum $\varphi := \varphi_1 + \varphi_2$ as an element of $W_0^{1,p}(\text{Int } A)$, and

$$\mathcal{F}(\ell_X + \varphi, \text{Int } A) = \sum_{i=1}^2 \mathcal{F}(\ell_X + \varphi_i, \text{Int } A_i).$$

  It follows

$$M(X; \mathcal{F}, A) \leq \sum_{i=1}^2 M(X; \mathcal{F}, A_i).$$

- Covariance follows immediately as

$$\begin{aligned}(\Upsilon_X(A) \circ \tau_z)(\mathcal{F}) &= M(X; \tau_z\mathcal{F}, \text{Int } A) \\ &= M(X; \mathcal{F}, z + \text{Int } A) \\ &= \Upsilon_X(z + A)(\mathcal{F}).\end{aligned}$$

- The spatial constant is obviously finite since it is non-negative.

The upper bound has been already proved. ∎

Now we may apply Theorem (4.2). It follows

Corollary (4.7). *Suppose* $\Phi : \Xi \to \mathscr{F}_\beta^p$ *is random integral functional that is periodic in law and ergodic. For every regular sequence of convex sets* $\{A_k\}_{k\in\mathbb{N}} \subset \mathcal{O}(\mathbb{R}^n)$ *with* $\lim_{k\to\infty} \rho(A_k) = \infty$, *there exists* $\Xi' \subset \Xi$ *with* $P(\Xi) = 1$ *such that*

$$\lim_{k\to\infty} \frac{1}{|A_k|} M(X; \Phi(\xi'); A_k) = \inf_{k\in\mathbb{N}} \frac{1}{k^n} \int_\Xi M(X; \Phi(\xi'), k\mathbb{I}^n)\, dP(\xi)$$

for every $\xi' \in \Xi'$ *and every* $X \in \mathbb{Q}^{m\times n}$.

Proof. By Theorem (4.2), for every $X \in \mathbb{Q}^{m\times n}$ there exists $\Xi_X' \subset \Xi$ with $P(\Xi_X') = 1$ such that the equality holds for all $\xi' \in \Xi_X'$. Now define $\Xi' := \bigcap_{X\in\mathbb{Q}^{m\times n}} \Xi_X'$. ∎

We supposed periodicity in law and ergodicity in this corollary. But the right-hand side makes sense for every random integral functional. Let us therefore for any family of functions $\{f(\xi) : \mathbb{R}^n \times \mathbb{R}^{m\times n} \to \mathbb{R}\}_{\xi\in\Xi}$ that induces a random integral functional on $\mathscr{F}_\beta^p$ define

$$f_{\text{hom}}(X) := \inf_{k\in\mathbb{N}} \frac{1}{k^n} \int_\Xi m\big(X; f(\xi), k\mathbb{I}^n\big)\, dP(\xi).$$

In particular, if the corresponding random integral functional is periodic in law and ergodic, then for almost every $\xi \in \Xi$ and for all $X \in \mathbb{Q}^{m\times n}$

$$\lim_{k\to\infty} \frac{1}{k^n} m\big(X; f(\xi); k\mathbb{I}^n\big) = f_{\text{hom}}(X).$$

Let us for further calculations prove the following approximation lemma.

Lemma (4.8). *Take any random integral functional* $\Phi : \Xi \to \mathscr{F}_\beta^p$ *and* $U \in \mathcal{O}(\mathbb{R}^n)$. *Let* $\{\varphi_j\}_{j\in\mathbb{N}} \subset W_0^{1,p}(U, \mathbb{R}^m)$ *be a dense subset. Then for every* $X \in \mathbb{R}^{m\times n}$

$$\int_\Xi M\big(X; \Phi(\xi), U\big)\, dP(\xi) = \inf\left\{ \sum_{j=1}^\infty \int_{\Xi_j} \Phi(\xi)\big(\ell_X + \varphi_j, U\big)\, dP(\xi) : \Xi = \bigsqcup_{j=1}^\infty \Xi_j \right\}.$$

More precisely, we assign to every φ_j *a measurable set* $\Xi_j \subset \Xi$ *such that* $\{\Xi_j\}_{j\in\mathbb{N}}$ *forms a measurable partition of* Ξ *and take infimum over all possible such choices.*

Proof. The right-hand side of the equality above is obviously not smaller than the left. Let us prove the other inequality.

Take arbitrary $\eta > 0$. For each $j \in \mathbb{N}$ the map

$$\Xi \to \mathbb{R}, \quad \xi \mapsto \Phi(\xi)\big(\ell_X + \varphi_j, U\big)$$

is measurable. Therefore, by $\Xi_0 := \emptyset$ and recursively

$$\Xi_j := \left\{ \xi \in \Xi : \Phi(\xi)(\ell_X + \varphi_j, U) < M\big(X; \Phi(\xi), U\big) + \eta \right\} \smallsetminus \bigcup_{i=0}^{j-1} \Xi_i$$

measurable disjoint subsets of Ξ are defined. Moreover, according to Lemma (4.5),

$$\bigcup_{j=1}^{\infty} \Xi_j = \bigcup_{j=1}^{\infty} \left\{ \xi \in \Xi : \Phi(\xi)(\ell_X + \varphi_j, U) < M\big(X; \Phi(\xi), U\big) + \eta \right\} = \Xi$$

(or at least up to a P-negligible set.) Hence, we obtained a measurable partition, and

$$\sum_{j=1}^{\infty} \int_{\Xi_j} \Phi(\xi)(\ell_X + \varphi_j, U) \, dP(\xi) \quad < \quad \sum_{j=1}^{\infty} \int_{\Xi_j} \Big(M\big(X; \Phi(\xi), U\big) + \eta \Big) \, dP(\xi)$$

$$= \int_{\Xi} M\big(X; \Phi(\xi), U\big) \, dP(\xi) + \eta.$$

The fact that η was arbitrary completes the proof. ∎

4.2 Stochastic homogenization

In Section 4.1 we defined the "homogenized" function without any real explanation or justification of the concept. For a random integral functional

$$\Phi : \Xi \to \mathscr{F}_\beta^p, \quad \Phi(\xi)(u, U) = \left\{ \begin{array}{ll} \int_U f(\xi)\big(x, \nabla u(x)\big) \, dx, & u \in W^{1,p}(\mathbb{R}^n; \mathbb{R}^m), \\ \infty, & \text{else,} \end{array} \right.$$

let us analogously to the deterministic case define for every $\varepsilon > 0$

$$\Phi_\varepsilon : \Xi \to \mathscr{F}_\beta^p, \quad \Phi_\varepsilon(\xi)(u, U) := \left\{ \begin{array}{ll} \int_U f(\xi)\big(\frac{x}{\varepsilon}, \nabla u(x)\big) \, dx, & u \in W^{1,p}(\mathbb{R}^n; \mathbb{R}^m), \\ \infty, & \text{else.} \end{array} \right.$$

Obviously, the image indeed lies in $\mathscr{F}_\beta^p$. This is also a random integral functional since for every $u \in W^{1,p}(\mathbb{R}^n, \mathbb{R}^m)$ and $U \in \mathcal{O}(\mathbb{R}^n)$ it holds

$$\Phi_\varepsilon(\xi)(u, U) = \int_U f(\xi) \left(\frac{x}{\varepsilon}, \nabla u(x) \right) \, dx$$

$$= \varepsilon^n \int_{\frac{1}{\varepsilon} U} f(\xi) \left(y, \frac{1}{\varepsilon} \nabla\big(u(\varepsilon y)\big) \right) \, dy$$

$$= \varepsilon^n \, \Phi(\xi) \left(\frac{1}{\varepsilon} u(\varepsilon \_), \frac{1}{\varepsilon} U \right).$$

Since also $\frac{1}{\varepsilon} u(\varepsilon \_) \in W^{1,p}(\mathbb{R}^n, \mathbb{R}^m)$ and $\frac{1}{\varepsilon} U \in \mathcal{O}(\mathbb{R}^n)$, the map $\xi \mapsto \Phi_\varepsilon(\xi)(u, U)$ is measurable, and therefore the map Φ_ε is also measurable. Moreover, let us define

$$\Phi_{\text{hom}}(u, U) := \left\{ \begin{array}{ll} \int_U f_{\text{hom}}(\nabla u(x)) \, dx, & u \in W^{1,p}(\mathbb{R}^n; \mathbb{R}^m), \\ \infty, & \text{else.} \end{array} \right.$$

In the deterministic setting, Φ_{hom} is the Γ-limit of the family $\{\Phi_\varepsilon\}_\varepsilon$. Earlier results for the stochastic case can be found in [DM:86-1, DM:86-2]. In both cases a more special case is considered with the inducing function being convex in the second variable. In the first article the authors do not suppose ergodicity, but require the random integral functionals to be in some sense independent. Furthermore, they also consider a family that has merely the same law as $\{\Phi_\varepsilon\}_\varepsilon$, which results in the convergence in probability and not in the almost sure convergence. In the second article ergodicity is assumed, and exactly the family $\{\Phi_\varepsilon\}_\varepsilon$ is considered. For more general results we refer to [MM:94]. We summarize Corollary 3.3 and Theorem 4.1 from that source in the following theorem.

Theorem (4.9). *Let*

$$\Phi : \Xi \to \mathscr{F}^p, \quad \Phi(\xi)(u,U) = \left\{ \begin{array}{ll} \int_U f(\xi)\big(x, \nabla u(x)\big) \ dx, & u \in W^{1,p}(\mathbb{R}^n; \mathbb{R}^m), \\ \infty, & else, \end{array} \right.$$

be a random integral functional. Suppose Φ is periodic in law and ergodic, and suppose there exist $\alpha, \beta, L > 0$ such that for almost every $\xi \in \Xi$

$$\alpha|X|^p \le f(\xi)(x,X) \le \beta\big(|X|^p + 1\big)$$

and

(4.10) $$|f(\xi)(x,X) - f(\xi)(x,Y)| \le L\big(1 + |X|^{p-1} + |Y|^{p-1}\big)|X - Y|.$$

Then there exists $\Xi' \in \mathcal{S}$ with $P(\Xi') = 1$ such that for every $\xi \in \Xi'$ and every $U \in \mathcal{A}(\mathbb{R}^n)$

$$\Phi_{\mathrm{hom}}(\_,U) = \Gamma(L^p)\text{-}\lim_{\varepsilon \to 0} \Phi_\varepsilon(\xi)(\_,U).$$

For the density it holds for every $\xi \in \Xi'$, every cube $Q \subset \mathbb{R}^n$ and every $X \in \mathbb{R}^{m \times n}$

$$f_{\mathrm{hom}}(X) = \lim_{t \to \infty} \frac{m\big(X; f(\xi), tQ\big)}{t^n|Q|^n}.$$

In [MM:94] the authors first show the last formula. Compared to the one in Corollary (4.7), the essential difference is that it now holds for all $X \in \mathbb{R}^{m \times n}$ and not only for $X \in \mathbb{Q}^{m \times n}$. For the transition they need the local Lipschitz continuity assumed in (4.10). Then they show the pivotal Γ-convergence result.

We wish to apply that kind of result without the condition (4.10) and with a weaker lower bound of Gårding type. The latter can be achieved similarly as for the deterministic case and will be explained later.

First, let us show that the Lipschitz condition can be circumvented. The idea is to pass to the quasiconvex envelope (in the second variable) since it, according to Lemma (A.4), automatically fulfils (4.10), and since the related infima coincide:

Lemma (4.11). *Suppose $f : \mathbb{R}^n \times \mathbb{R}^{m \times n} \to \mathbb{R}$ is a Carathéodory function satisfying*

$$0 \le f(x,X) \le \beta(|X|^p + 1)$$

for almost all $x \in \mathbb{R}^n$ and $X \in \mathbb{R}^{m \times n}$. Then for every $U \in \mathcal{O}(\mathbb{R}^n)$ and every $X \in \mathbb{R}^{m \times n}$

$$m(X; f, U) = m(X; f^{\mathrm{qc}}, U).$$

For the proof we refer to, e.g., Theorem 9.8 in [Dac:08].

The same concept was employed in Section 3 of [BF:07] for a deterministic (but more general) case. Now the following problem arises: If we have a random integral functional and replace the underlying function by its quasiconvex envelope, is the new mapping still a random integral functional? We especially have to look into its measurability.

Henceforth we suppose $p \neq 1$.

Let us for the start take arbitrary functional $\mathcal{F} \in \mathscr{F}^p_{\alpha,\beta}$ determined by the function f. By replacing f with f^{qc}, we get the lower semicontinuous envelope of the functional, as seen in Theorem (B.5). More precisely, by defining

$$(\mathrm{lsc}\,\mathcal{F})(u, U) := \begin{cases} \int_U f^{qc}(x, \nabla u(x))\ dx, & u \in W^{1,p}(\mathbb{R}^n; \mathbb{R}^m), \\ \infty, & \text{else.} \end{cases}$$

we again get an element in $\mathscr{F}^p_{\alpha,\beta}$ and $(\mathrm{lsc}\,\mathcal{F})(\_, U) = \mathrm{lsc}(\mathcal{F}(\_, U))$ for every $U \in \mathcal{A}(\mathbb{R}^n)$. Choose $u \in W^{1,p}(\mathbb{R}^n; \mathbb{R}^m)$ and $U \in \mathcal{A}(\mathbb{R}^n)$. We can express the lower semicontinuous envelope by the Yosida transform (e.g., Proposition 8.1 in [BD:98]):

$$(\mathrm{lsc}\,\mathcal{F})(u, U) = \sup_{\lambda \geq 0} \inf_{v \in L^p(\mathbb{R}^n, \mathbb{R}^m)} \left(\mathcal{F}(u + v, U) + \lambda \|v\|^p_{L^p(U, \mathbb{R}^m)} \right).$$

Knowing the domain of $\mathcal{F}$ and $\mathrm{lsc}\,\mathcal{F}$, we may write

$$(\mathrm{lsc}\,\mathcal{F})(u, U) = \sup_{\lambda \geq 0} \inf_{v \in W^{1,p}(\mathbb{R}^n, \mathbb{R}^m)} \left(\mathcal{F}(u + v, U) + \lambda \|v\|^p_{L^p(U, \mathbb{R}^m)} \right).$$

We notice that the expression that we are maximizing is non-decreasing in λ, so we may restrict ourselves to $\lambda \in \mathbb{N}$. Moreover, if we choose a dense subset $\{v_j\}_{j \in \mathbb{N}} \subset W^{1,p}(\mathbb{R}^n, \mathbb{R}^m)$, then by applying Lemma (4.5)

$$(\mathrm{lsc}\,\mathcal{F})(u, U) = \sup_{k \in \mathbb{N}} \inf_{j \in \mathbb{N}} \left(\mathcal{F}(u + v_j, U) + k \|v_j\|^p_{L^p(U, \mathbb{R}^m)} \right).$$

(Note that this holds for every $\mathcal{F} \in \mathscr{F}^p_{\alpha,\beta}$.) By the definition of the σ-algebra on $\mathscr{F}^p$, each map

$$\mathscr{F}^p_{\alpha,\beta} \to \mathbb{R}, \quad \mathcal{F} \mapsto \mathcal{F}(u + v_j, U)$$

is measurable. Therefore, the map

$$\mathscr{F}^p_{\alpha,\beta} \to \mathbb{R}, \quad \mathcal{F} \mapsto (\mathrm{lsc}\,\mathcal{F})(u, U)$$

is measurable as well, since in the Yosida transform description above we have countably many measurable functions.

If $U \in \mathcal{O}(\mathbb{R}^n) \setminus \mathcal{A}(\mathbb{R}^n)$, we may find a non-decreasing sequence $\{U_i\}_{i \in \mathbb{N}} \subset \mathcal{A}(U)$ with $|U \setminus U_i| \to 0$. Then

$$(\mathrm{lsc}\,\mathcal{F})(u, U) = \sup_{i \in \mathbb{N}} (\mathrm{lsc}\,\mathcal{F})(u, U_i)$$

for every $\mathcal{F} \in \mathscr{F}^p_{\alpha,\beta}$. Therefore, $\mathcal{F} \mapsto (\mathrm{lsc}\,\mathcal{F})(u, U)$ is measurable for all $u \in W^{1,p}(\mathbb{R}^n; \mathbb{R}^m)$ and all $U \in \mathcal{O}(\mathbb{R}^n)$. Hence, the mapping

$$\mathscr{F}^p_{\alpha,\beta} \to \mathscr{F}^p_{\alpha,\beta}, \quad \mathcal{F} \mapsto \mathrm{lsc}\,\mathcal{F}$$

is measurable. It follows

Lemma (4.12). *If $\Phi : \Xi \to \mathscr{F}^p_{\alpha,\beta}$ is a random integral functional, then so is*

$$\text{lsc } \Phi : \Xi \to \mathscr{F}^p_{\alpha,\beta}, \quad \xi \mapsto \text{lsc } \Phi(\xi).$$

Proof. Since the map $\mathcal{F} \mapsto \text{lsc } \mathcal{F}$ is measurable, $\text{lsc } \Phi$ is a composition of two measurable maps. ∎

Since we quasiconvexify in the second variable, the first being fixed, it holds

$$\tau_z(\text{lsc } \mathcal{F}) = \text{lsc}(\tau_z \mathcal{F})$$

for every $\mathcal{F} \in \mathscr{F}^p_{\alpha,\beta}$ and $z \in \mathbb{Z}^n$.

Corollary (4.13). *Theorem (4.9) holds without the assumption of local Lipschitz continuity.*

Proof. For a given family $\{f(\xi) : \mathbb{R}^n \times \mathbb{R}^{m \times n} \to \mathbb{R}\}_{\xi \in \Xi}$, the family $\{f(\xi)^{\text{qc}}\}_{\xi \in \Xi}$ fulfils the same growth conditions and condition (4.10). The mapping $\text{lsc } \Phi : \Xi \to \mathscr{F}^p_{\alpha,\beta}$ is also

- periodic in law:

  For every $E \in \mathscr{F}^p_{\alpha,\beta}$ it holds

  $$\begin{aligned}
  P(\tau_z(\text{lsc } \Phi) \in E) &= P(\text{lsc}(\tau_z \Phi) \in E) \\
  &= P(\tau_z \Phi \in \text{lsc}^{-1}(E)) \\
  &= P(\Phi \in \text{lsc}^{-1}(E)) \\
  &= P(\text{lsc } \Phi \in E).
  \end{aligned}$$

- ergodic:

  Suppose $\tau_z(E) = E$ for every $z \in \mathbb{Z}^n$. Then $\tau_z(\text{lsc}^{-1}(E)) = \text{lsc}^{-1}(\tau_z(E)) = \text{lsc}^{-1}(E)$ for every $z \in \mathbb{Z}^n$ as well. It follows

  $$P(\text{lsc } \Phi \in E) = P(\Phi \in \text{lsc}^{-1}(E)) \in \{0,1\}.$$

According to Theorem (4.9), there exists $\Xi' \in \mathcal{S}$ with $P(\Xi') = 1$ such that for every $\xi \in \Xi'$ and every $U \in \mathcal{A}(\mathbb{R}^n)$

$$\begin{aligned}
(\text{lsc } \Phi)_{\text{hom}}(\_,U) &= \Gamma(L^p)\text{-}\lim_{\varepsilon \to 0} \big(\text{lsc } \Phi_\varepsilon(\xi)\big)(\_,U) \\
&= \Gamma(L^p)\text{-}\lim_{\varepsilon \to 0} \Phi_\varepsilon(\xi)(\_,U).
\end{aligned}$$

The density of $(\text{lsc } \Phi)_{\text{hom}}$ is $(f^{\text{qc}})_{\text{hom}}$, where f^{qc} stands for the corresponding family. From Lemma (4.11) it follows $(f^{\text{qc}})_{\text{hom}} = f_{\text{hom}}$, and thus also $(\text{lsc } \Phi)_{\text{hom}} = \Phi_{\text{hom}}$. ∎

Now we may, as in the deterministic case, generalize this results to random integral functionals of Gårding type. For that purpose, for $p > 1$, $\beta > 0$ and a lower bound of p-Gårding type g (as in Definition (2.28)), we denote by $\mathscr{G}^p_{g,\beta}$ the set integral functionals from of $\mathscr{F}^p_\beta$ whose densities additionally fulfil

$$f(x,X) \geq g(X)$$

for a.e. $x \in \mathbb{R}$ and all $X \in \mathbb{R}^{m \times n}$.

Proposition (4.14). *Suppose that* $\Phi : \Xi \to \mathscr{G}^p_{g,\beta}$ *is a random integral functional that is periodic in law and ergodic. Then for P-a.e. $\xi \in \Xi$ and all $U \in \mathcal{A}(\mathbb{R}^n)$,*

$$\Phi_{\text{hom}}(\_, U) = \Gamma(L^p)\text{-}\lim_{\varepsilon \to 0} \Phi_{\varepsilon}(\xi)(\_, U).$$

The density fulfils for P-a.e. $\xi \in \Xi$, every $U \in \mathcal{A}(\mathbb{R}^n)$ and every $X \in \mathbb{R}^{m \times n}$

$$f_{\text{hom}}(X) = \lim_{t \to \infty} \frac{m(X; f(\xi), tU)}{t^n |U|^n}.$$

Proof. Let us define random integral functionals $\Phi^{(j)} : \Xi \to \mathscr{F}^p_{\beta+1}$ by setting their densities to be

$$f^{(j)}(\xi)(x, X) := f(\xi)(x, X) + \tfrac{1}{j}|X|^p.$$

(Measurability is obvious.) If we denote by $\mathcal{F}_0$ the integral functional

$$\mathcal{F}_0(u, U) := \begin{cases} \int_U |\nabla u(x)|^p \, dx, & u \in W^{1,p}(\mathbb{R}^n; \mathbb{R}^m), \\ \infty, & \text{else}, \end{cases}$$

then simply $\Phi^{(j)} = \Phi + \tfrac{1}{j}\mathcal{F}_0$. Since $\tau_z \mathcal{F}_0 = \mathcal{F}_0$ for all $z \in \mathbb{Z}^n$, each $\Phi^{(j)} \in \mathscr{F}^p_{1/j,\beta+1/j}$ is again periodic in law and ergodic. By Corollary (4.13) there exists $\Xi' \subset \Xi$ with $P(\Xi') = 1$ such that for all $j \in \mathbb{N}$, $\xi \in \Xi'$ and $U \in \mathcal{A}(\mathbb{R}^n)$

$$\Gamma(L^p)\text{-}\lim_{\varepsilon \to 0} \Phi^{(j)}_{\varepsilon}(\xi)(\_, U) = \Phi^{(j)}_{\text{hom}}(\_, U).$$

Hence, for any fixed $\xi \in \Xi'$, the function $f^{(j)}(\xi)$ is homogenizable for every $j \in \mathbb{N}$, and we have the deterministic case from Theorem (2.31):

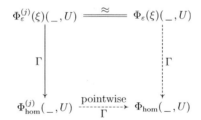

The representation formula follows from Proposition (2.29). ∎

In the deterministic case the standard result considers the integral functionals whose densities are $\mathbb{I}^n$-periodic in the first variable. However, the homogenization closure result holds for all integral functionals of Gårding type with homogenizable densities. The role of periodicity now play periodicity in law and ergodicity (together), but we can do the same generalization also in the stochastic setting.

Definition (4.15). *A random integral functional $\Phi : \Xi \to \mathscr{F}^p$ is homogenizable if there exists $\mathcal{F} \in \mathscr{F}^p$ with location-independent density such that for P-a.e. $\xi \in \Xi$ and every $U \in \mathcal{A}(\mathbb{R}^n)$*

$$\Gamma(L^p)\text{-}\lim_{\varepsilon \to 0} \Phi_{\varepsilon}(\xi)(\_, U) = \mathcal{F}(\_, U).$$

From the representation formula in Proposition (2.29), it follows that every homogenizable random integral functional $\Phi : \Xi \to \mathscr{G}^p_{g,\beta}$ almost surely Γ-converges to $\mathcal{F} = \Phi_{\text{hom}}$ (with density f_{hom}).

Theorem (4.16). *Let us have random integral functionals*

$$\Phi^{(j)} : \Xi \to \mathscr{G}^p_{g^{(j)},\beta}, \quad j \in \mathbb{N} \cup \{\infty\},$$

with densities $\{f^{(j)}(\xi)\}_{\xi\in\Xi}$. Suppose that

- $g^{(j)}$, $j \in \mathbb{N} \cup \{\infty\}$, *is a family of lower bounds of uniform p-Gårding type,*

- *for every $j \in \mathbb{N}$ the random integral functional $\Phi^{(j)}$ is homogenizable,*

- *for P-a.e. $\xi \in \Xi$ and every $R > 0$*

$$\lim_{j\to\infty} \limsup_{T\to\infty} \frac{1}{(2T)^n} \int_{(-T,T)^n} \sup_{|X|\le R} |f^{(j)}(\xi)(x,X) - f^{(\infty)}(\xi)(x,X)|\, dx = 0.$$

Then also $\Phi^{(\infty)}$ is homogenizable and

$$f^{(\infty)}_{\text{hom}}(X) = \lim_{j\to\infty} f^{(j)}_{\text{hom}}(X).$$

Proof. By the assumptions, there exists $\Xi' \subset \Xi$ with $P(\Xi') = 1$ such that for every $\xi \in \Xi'$

- the equivalence assumption holds,

- $\Phi^{(j)}(\xi)$ is homogenizable for each $j \in \mathbb{N}$.

For arbitrary fixed $\xi \in \Xi'$, we may apply Theorem (2.31).   ∎

For the sake of completeness, we also adapt Theorem (2.34).

Theorem (4.17). *Suppose we have the setting from Theorem (4.16), and moreover, suppose*

$$\lim_{j\to\infty} \int_{(-T,T)^n} \sup_{|X|\le R} |f^{(j)}(\xi)(x,X) - f^{(\infty)}(\xi)(x,X)|\, dx = 0$$

for P-a.e. $\xi \in \Xi$ and all $T, R > 0$. Then the following diagram commutes for P-a.e. $\xi \in \Xi$ and every $U \in \mathcal{A}(\mathbb{R}^n)$:

$$
\begin{array}{ccc}
\Phi^{(j)}_\varepsilon(\xi)(\_,U) & \xrightarrow{\ \Gamma\ } & \operatorname{lsc} \Phi^{(\infty)}_\varepsilon(\xi)(\_,U) \\
\Big\downarrow{\scriptstyle\Gamma} & & \Big\downarrow{\scriptstyle\Gamma} \\
\Phi^{(j)}_{\text{hom}}(\_,U) & \xrightarrow{\ \Gamma\ } & \Phi^{(\infty)}_{\text{hom}}(\_,U)
\end{array}
$$

Remark (4.18). Let us at the end mention that it is possible to apply the derived results to elasticity theory, mutatis mutandis, as we did in Chapter 3 with the results from Chapter 2. For one-well stored-energy functions, the commutability of linearization and stochastic homogenization was proved in [GN:11].

Chapter 5

The $p = 1$ case

The results in Chapters 2 and 4 for $p > 1$ lead to a natural question if the same holds for $p = 1$. The answer is negative, as we will show with an explicit counterexample, the problem being that in $W^{1,1}$ boundedness of gradients does not imply their equiintegrability. As already known from the literature, in this case the behaviour at ∞, described by the recession function, also must be taken into account. Therefore, for analogous commutability results an additional assumption is needed.

Since our interest is purely theoretic, we will not try to find a counterpart to all preceding results. We will only deal with the most basic deterministic setting from Theorem (2.2), since it already contains the most important ideas.

5.1 Counterexample

The counterexample is based on Example 6.4 in [BD:93]. We will consider the scalar case $m = 1$ with $\Omega = J := (-1, 1)$ and show that even the perturbation result from Theorem (2.22) does not hold.

For $p = 1$, integral functionals with convex densities are not lower semicontinuous. The relaxed functionals have an extra term determined by the recession function. We will postpone a more detailed discussion about recession functions until Section 8.3. Until then it suffices that for a convex function $f : \mathbb{R}^n \to [0, \infty)$ with a linear upper bound it may be computed by

$$f^\infty(\xi) = \lim_{t \to \infty} \frac{f(t\xi)}{t}.$$

Let us take the convex function

$$f : \mathbb{R} \to \mathbb{R}, \quad f(\xi) := \max\{|\xi|, 2|\xi| - 1\}.$$

Clearly, $f^\infty(\xi) = \lim_{t \to \infty} \frac{f(t\xi)}{t} = 2|\xi|$.

As the first family of functionals, we choose a constant family given by

$$\mathcal{F}(u) := \int_{-1}^{1} f(u'(x)) \, dx \qquad \text{for } u \in W^{1,1}(J)$$

and ∞ otherwise on $L^1(J)$. It $\Gamma(L^1)$-converges to its relaxation $\mathcal{F}_0$, which has domain $BV(J)$ and there takes the values

$$
\begin{aligned}
\mathcal{F}_0(u) &= \int_{-1}^{1} f(u'(x)) \, dx + \int_{-1}^{1} f^\infty \left(\tfrac{d(D^s u)}{d|D^s u|}(x) \right) d|D^s u|(x) \\
&= \int_{-1}^{1} f(u'(x)) \, dx + 2\|D^s u\|_{M(J)},
\end{aligned}
$$

see, e.g., Theorem 4.1 in [AD:92]. Here u' stands for the Radon-Nikodym derivative of Du with respect to the Lebesgue measure $\mathcal{L}^1$. For the second family let

$$
a_\varepsilon(x) := \left\{ \begin{array}{ll} 1, & |x| \geq \varepsilon, \\ \frac{1}{2\varepsilon}, & |x| < \varepsilon, \end{array} \right.
$$

We define densities $g_\varepsilon : J \times \mathbb{R} \to \mathbb{R}$ by

$$
g_\varepsilon(x, \xi) := f \left(\tfrac{\xi}{a_\varepsilon(x)} \right) a_\varepsilon(x) = \left\{ \begin{array}{ll} f(\xi), & |x| \geq \varepsilon, \\ \frac{1}{2\varepsilon} f(2\varepsilon\xi), & |x| < \varepsilon. \end{array} \right.
$$

Then the (one-index) family $\{g_\varepsilon\}_{\varepsilon > 0}$ and the constant one given by f are equivalent since for any $R > 0$

$$
\begin{aligned}
\limsup_{\varepsilon \to 0} \int_{-1}^{1} \sup_{|\xi| \leq R} |g_\varepsilon(x, \xi) - f(\xi)| \, dx &= \limsup_{\varepsilon \to 0} \int_{-\varepsilon}^{\varepsilon} \sup_{|\xi| \leq R} |\tfrac{1}{2\varepsilon} f(2\varepsilon\xi) - f(\xi)| \, dx \\
&\leq \limsup_{\varepsilon \to 0} \int_{-\varepsilon}^{\varepsilon} \sup_{|\xi| \leq R} 4|\xi| \, dx \\
&= \limsup_{\varepsilon \to 0} 8R\varepsilon \\
&= 0.
\end{aligned}
$$

Let us denote $\lambda_\varepsilon := a_\varepsilon \mathcal{L}^1 \in M(J)$. For $u \in W^{1,1}(J)$ it is

$$
Du = u' \mathcal{L}^1 = \frac{u'}{a_\varepsilon} \lambda_\varepsilon.
$$

Hence, the corresponding functionals $\mathcal{G}_\varepsilon$ have on $W^{1,1}(J)$ the following representation

$$
\mathcal{G}_\varepsilon(u) = \int_{-1}^{1} g_\varepsilon(x, u'(x)) \, dx = \int_{-1}^{1} f \left(\tfrac{dDu}{d\lambda_\varepsilon}(x) \right) d\lambda_\varepsilon(x).
$$

Since

$$
\lambda_\varepsilon \overset{*}{\rightharpoonup} \lambda := \delta_0 + \mathcal{L}^1 \quad \text{in } M(J),
$$

it follows by the results in [BF:91] that $\mathcal{G}_\varepsilon \ \Gamma(L^1)$-converges to $\mathcal{G}_0$ where

$$
\mathcal{G}_0(u) = \int_{-1}^{1} f \left(\tfrac{d(D_\lambda^a u)}{d\lambda}(x) \right) d\lambda(x) + \int_{-1}^{1} f^\infty \left(\tfrac{d(D_\lambda^s u)}{d|D_\lambda^s u|}(x) \right) d|D_\lambda^s u|(x)
$$

if $u \in BV(J)$ and ∞ otherwise. $D_\lambda^a u$ and $D_\lambda^s u$ in the formula stand for the absolutely continuous resp. singular part of Du with respect to λ. We compare these two decompositions

$$
Du = u' \mathcal{L}^1 + D^s u = \tfrac{d(D_\lambda^a u)}{d\lambda} (\mathcal{L}^1 + \delta_0) + D_\lambda^s u
$$

and arrive at

$$\frac{d(D_\lambda^a u)}{d\lambda} = u' \quad \mathcal{L}^1\text{-a.e.}, \quad \frac{d(D_\lambda^a u)}{d\lambda}(0) = Du(\{0\}) \quad \text{and} \quad (D^s u)\lfloor_{J\setminus\{0\}} = (D_\lambda^s u)\lfloor_{J\setminus\{0\}}.$$

Therefore

$$\mathcal{G}_0(u) = \int_{-1}^1 f(u'(x)) \, dx + f(Du(\{0\})) + 2\|D_\lambda^s u\|_{M(J)}.$$

Hence,

$$\mathcal{F}_0(u) = \int_{-1}^1 f(u'(x)) \, dx + 2\|D^s u\|_{M(J)},$$

$$\mathcal{G}_0(u) = \int_{-1}^1 f(u'(x)) \, dx + f(Du(\{0\})) + 2\|D_\lambda^s u\|_{M(J)}.$$

If we choose $u := 1_{(0,1)}$, we have $Du = \delta_0$ and

$$u' = 0 \quad \mathcal{L}^1\text{-a.e.}, \quad D^s u = \delta_0, \quad D_\lambda^a u = \delta_0, \quad D_\lambda^s u = 0.$$

Hence

$$\mathcal{F}_0(1_{(0,1)}) = 2 \neq 1 = \mathcal{G}_0(1_{(0,1)}).$$

This example shows also another important issue (which is the reason of the discussion in [BD:93]). In the case $p > 1$ the limiting functional even has a density, i.e., it is given by some Borel function $\varphi : \Omega \times \mathbb{R}^{m\times n} \to \mathbb{R}$ with the same growth properties such that the values on $W^{1,p}(\Omega; \mathbb{R}^m)$ are given by $\int_\Omega \varphi(x, \nabla u(x)) \, dx$ (and ∞ elsewhere). For $p = 1$ it is still true that there is such a density that determines the values on $W^{1,1}(\Omega; \mathbb{R}^m)$ (see Theorem 12.5 in [BD:98] or Teorema in [DG:75]). However, it is in general wrong that the values on $BV(\Omega; \mathbb{R}^m)$ are given by

$$\int_\Omega \varphi(x, \nabla u(x)) \, dx + \int_\Omega \varphi^\infty \left(x, \frac{d(D^s u)}{d|D^s u|}(x) \right) d|D^s u|(x).$$

Indeed, in the example above the density on $W^{1,1}(J)$ is in both cases f. However, the formula above yields $\mathcal{F}_0$. The limiting functional $\mathcal{G}_0$ for the family $\mathcal{G}_\varepsilon$ has a different structure.

5.2 Γ-closure

The counterexample in Section 5.1 indicates that the problem occurs when dealing with families whose difference grows linearly for large X at least on some set of x. The equivalence condition from Definition (2.1) does not exclude such behaviour. Therefore, let us additionally introduce a notion of equivalence of families at large arguments.

Definition (5.1). Let $\Omega \subset \mathbb{R}^n$ be open, and suppose

$$f_\varepsilon^{(j)} : \Omega \times \mathbb{R}^{m\times n} \to \mathbb{R}, \quad j \in \mathbb{N} \cup \{\infty\}, \ \varepsilon > 0,$$

are Borel functions. We say that the families $\left\{ \{f_\varepsilon^{(j)}\}_{\varepsilon>0} \right\}_{j\in\mathbb{N}}$ and $\{f_\varepsilon^{(\infty)}\}_{\varepsilon>0}$ *are equivalent at* ∞ if for

$$r_\varepsilon^{(j)}(R) := \operatorname*{ess\,sup}_{x\in\Omega} \sup_{|X|\geq R} \frac{|f_\varepsilon^{(j)}(x, X) - f_\varepsilon^{(\infty)}(x, X)|}{|X|}$$

it holds
$$\lim_{R \to \infty} \limsup_{j \to \infty} \limsup_{\varepsilon \to 0} r_\varepsilon^{(j)}(R) = 0.$$

Remark (5.2).

(a) The equivalence condition from Definition (2.1) was a sort of combination of the local uniform convergence in X and the L^1-convergence in x. Here we have a uniform convergence of the slopes for large X and the L^∞-convergence in x.

(b) This assumption is met if for some $\delta \in (0, 1)$ and $\gamma > 0$ it holds for all $j \in \mathbb{N}$ and $\varepsilon > 0$
$$|f_\varepsilon^{(j)}(x, X) - f_\varepsilon^{(\infty)}(x, X)| \le \gamma |X|^{1-\delta}$$
for a.e. $x \in \Omega$ and $X \in \mathbb{R}^{m \times n}$. An analogous condition was already introduced when dealing with the relaxation for $p = 1$, see, e.g. Section 3 in [BFT:00] and Section 4 in [BFM:98].

(c) Although the recession function plays an important role in the relaxation, it does not suffice to impose just their convergence in some sense. Bear in mind that in our counterexample they even coincide, but still the statement does not hold.

Theorem (5.3). *Let $\Omega \subset \mathbb{R}^n$ be bounded and open. Suppose that the family of Borel functions $f_\varepsilon^{(j)} : \Omega \times \mathbb{R}^{m \times n} \to \mathbb{R}$, $j \in \mathbb{N} \cup \{\infty\}$, $\varepsilon > 0$, uniformly fulfils the standard linear growth condition, and define*

$$\mathcal{F}_\varepsilon^{(j)}(u) := \begin{cases} \int_\Omega f_\varepsilon^{(j)}(x, \nabla u(x)) \, dx, & u \in W^{1,1}(\Omega; \mathbb{R}^m), \\ \infty, & else. \end{cases}$$

Assume that

- *for each $j \in \mathbb{N}$ the Γ-limit $\Gamma(L^1)$-$\lim_{\varepsilon \to 0} \mathcal{F}_\varepsilon^{(j)} =: \mathcal{F}_0^{(j)}$ exists,*

- *the families $\{\{f_\varepsilon^{(j)}\}_{\varepsilon > 0}\}_{j \in \mathbb{N}}$ and $\{f_\varepsilon^{(\infty)}\}_{\varepsilon > 0}$ are equivalent and also equivalent at ∞ on Ω.*

Then also $\Gamma(L^1)$-$\lim_{\varepsilon \to 0} \mathcal{F}_\varepsilon^{(\infty)} =: \mathcal{F}_0^{(\infty)}$ exists. It is the pointwise and the Γ-limit of $\mathcal{F}_0^{(j)}$ as $j \to \infty$:

$$\mathcal{F}_0^{(\infty)} = \lim_{j \to \infty} \mathcal{F}_0^{(j)} = \Gamma(L^1)\text{-} \lim_{j \to \infty} \mathcal{F}_0^{(j)}.$$

Before proving, let us state a compactness result for this setting. For its proof, consult Section 6 in [BD:93] and the references therein.

Lemma (5.4). *Let us have a sequence of Borel functions $g_k : \Omega \times \mathbb{R}^{m \times n} \to \mathbb{R}$, $k \in \mathbb{N}$, for which there exist $\alpha, \beta > 0$ such that*

$$\alpha |X| \le g_k(x, X) \le \beta(|X| + 1)$$

for a.e. $x \in \Omega$ and all $X \in \mathbb{R}^{m \times n}$. For the sequence of integral functionals $\mathcal{G}_k$, $k \in \mathbb{N}$, on $L^1(\Omega; \mathbb{R}^m) \times \mathcal{A}(\Omega)$ given by

$$\mathcal{G}_k(u, U) := \begin{cases} \int_U g_k(x, \nabla u(x)) \, dx, & u \in W^{1,1}(U; \mathbb{R}^m), \\ \infty, & else, \end{cases}$$

there exists a subsequence $\{\mathcal{G}_{k_i}\}_{i\in\mathbb{N}}$ and a functional $\mathcal{G} : L^1(\Omega;\mathbb{R}^m) \times \mathcal{A}(\Omega) \to \mathbb{R} \cup \{\infty\}$ such that

$$\Gamma(L^1)\text{-}\lim_{i\to\infty}\mathcal{G}_{k_i}(\_,U) = \mathcal{G}(\_,U)$$

for every $U \in \mathcal{A}(\Omega)$. Moreover,

$$\alpha|Du|(U) \leq \mathcal{G}(u,U) \leq \beta(|U| + |Du|(U))$$

with the convention $|Du|(U) = \infty$ for $u \notin BV(U;\mathbb{R}^m)$.

Remark (5.5). The single domain version of Lemma (5.4), which is needed in the proof of Theorem (5.3), can be derived directly. However, we stated Lemma (5.4) for a possible extension of the result to variable domains.

For the proof of Theorem (5.3), we use the same strategy as in Theorem (2.2) incorporating the additional assumption on the behaviour for large X.

Proof (of Theorem (5.3)). First we assume that

$$\Gamma(L^1)\text{-}\lim_{\varepsilon\to 0}\mathcal{F}_\varepsilon^{(\infty)} =: \mathcal{F}_0^{(\infty)}$$

exists and that $\mathcal{F}_0^{(\infty)}(u) < \infty$ if and only if $u \in BV(\Omega;\mathbb{R}^m)$. This will be justified in Step 3. For the sake of simplicity, we suppose that all the densities are non-negative, i.e., they fulfil

$$\alpha|X| \leq f_\varepsilon^{(j)}(x,X) \leq \beta(|X| + 1).$$

Step 1: Upper bound. For $u \in L^1(\Omega;\mathbb{R}^m)$ we claim that

$$\limsup_{j\to\infty}\mathcal{F}_0^{(j)}(u) \leq \mathcal{F}_0^{(\infty)}(u).$$

By our assumption, this is obvious if $u \in L^1(\Omega;\mathbb{R}^m) \setminus BV(\Omega;\mathbb{R}^m)$. Let us take arbitrary $u \in BV(\Omega;\mathbb{R}^m)$. Fix any $\eta > 0$, and choose a recovery sequence $\{u_\varepsilon\}_\varepsilon \subset L^1(\Omega;\mathbb{R}^m)$ for u, i.e., $\lim_{\varepsilon\to 0}\mathcal{F}_\varepsilon^{(\infty)}(u_\varepsilon) = \mathcal{F}_0^{(\infty)}(u)$. We may suppose that $\{\mathcal{F}_\varepsilon^{(\infty)}(u_\varepsilon)\}_\varepsilon$ is bounded and consequently $\{u_\varepsilon\}_\varepsilon \subset W^{1,1}(\Omega;\mathbb{R}^m)$ with $\sup_\varepsilon \|\nabla u_\varepsilon\|_{L^1} =: B < \infty$. We choose $M > 0$ such that

$$\limsup_{j\to\infty}\ \limsup_{\varepsilon\to 0}\ r_\varepsilon^{(j)}(M) < \frac{\eta}{B},$$

$r_\varepsilon^{(j)}$ being as in Definition (5.1), and define the sets $E_\varepsilon := \{x \in \Omega : |\nabla u_\varepsilon(x)| \geq M\}$. After splitting

$$\mathcal{F}_\varepsilon^{(\infty)}(u_\varepsilon) = \int_{\Omega\setminus E_\varepsilon} f_\varepsilon^{(\infty)}(x,\nabla u_\varepsilon(x))\ dx + \int_{E_\varepsilon} f_\varepsilon^{(\infty)}(x,\nabla u_\varepsilon(x))\ dx,$$

we bound both terms from below by using $f_\varepsilon^{(j)}$. Obviously

$$\int_{\Omega\setminus E_\varepsilon} f_\varepsilon^{(\infty)}(x,\nabla u_\varepsilon(x))\ dx \geq \int_{\Omega\setminus E_\varepsilon} f_\varepsilon^{(j)}(x,\nabla u_\varepsilon(x))\ dx - \int_\Omega \sup_{|X|\leq M} |f_\varepsilon^{(j)}(x,X) - f_\varepsilon^{(\infty)}(x,X)|\ dx,$$

and furthermore

$$
\begin{aligned}
\int_{E_\varepsilon} & f_\varepsilon^{(\infty)}(x, \nabla u_\varepsilon(x)) \, dx \\
&\geq \int_{E_\varepsilon} f_\varepsilon^{(j)}(x, \nabla u_\varepsilon(x)) \, dx - \int_{E_\varepsilon} |f_\varepsilon^{(j)}(x, \nabla u_\varepsilon(x)) - f_\varepsilon^{(\infty)}(x, \nabla u_\varepsilon(x))| \, dx \\
&\geq \int_{E_\varepsilon} f_\varepsilon^{(j)}(x, \nabla u_\varepsilon(x)) \, dx - \int_{E_\varepsilon} r_\varepsilon^{(j)}(M) |\nabla u_\varepsilon(x)| \, dx \\
&\geq \int_{E_\varepsilon} f_\varepsilon^{(j)}(x, \nabla u_\varepsilon(x)) \, dx - B \, r_\varepsilon^{(j)}(M).
\end{aligned}
$$

Hence,

$$
\mathcal{F}_\varepsilon^{(\infty)}(u_\varepsilon) \geq \mathcal{F}_\varepsilon^{(j)}(u_\varepsilon) - \int_\Omega \sup_{|X| \leq M} |f_\varepsilon^{(j)}(x, X) - f_\varepsilon^{(\infty)}(x, X)| \, dx - B \, r_\varepsilon^{(j)}(M).
$$

Taking lim inf as $\varepsilon \to 0$ and employing the lim inf-inequality for $\Gamma(L^1)\text{-}\lim_{\varepsilon \to 0} \mathcal{F}_\varepsilon^{(j)} = \mathcal{F}_0^{(j)}$ yields

$$
\mathcal{F}_0^{(\infty)}(u) \geq \mathcal{F}_0^{(j)}(u) - \limsup_{\varepsilon \to 0} \int_\Omega \sup_{|X| \leq M} |f_\varepsilon^{(j)}(x, X) - f_\varepsilon^{(\infty)}(x, X)| \, dx - \limsup_{\varepsilon \to 0} B \, r_\varepsilon^{(j)}(M).
$$

By sending also $j \to \infty$, we arrive at

$$
\mathcal{F}_0^{(\infty)}(u) \geq \limsup_{j \to \infty} \mathcal{F}_0^{(j)}(u) - \limsup_{j \to \infty} \limsup_{\varepsilon \to 0} \int_\Omega \sup_{|X| \leq M} |f_\varepsilon^{(j)}(x, X) - f_\varepsilon^{(\infty)}(x, X)| \, dx - \eta.
$$

The claim now follows from the equivalence of the families and the arbitrariness of η.

Step 2: Lower bound. We claim that

$$
\liminf_{j \to \infty} \mathcal{F}_0^{(j)}(u_j) \geq \mathcal{F}_0^{(\infty)}(u)
$$

whenever $u_j \to u$ in $L^1(\Omega; \mathbb{R}^m)$.

Suppose

$$
u_j \to u \quad \text{in } L^1(\Omega; \mathbb{R}^m) \quad \text{and} \quad \liminf_{j \to \infty} \mathcal{F}_0^{(j)}(u_j) < \infty.
$$

From

$$
\lim_{R \to \infty} \limsup_{j \to \infty} \limsup_{\varepsilon \to 0} r_\varepsilon^{(j)}(R) = 0,
$$

$r_\varepsilon^{(j)}(R)$ being as in Definition (5.1), it follows that there is an increasing unbounded sequence $\{M_k\}_{k \in \mathbb{N}}$ such that

$$
\limsup_{j \to \infty} \limsup_{\varepsilon \to 0} r_\varepsilon^{(j)}(M_k) \leq \tfrac{1}{k}.
$$

We may find a subsequence $\{j_k\}_{k \in \mathbb{N}}$ such that

- all $\mathcal{F}_0^{(j_k)}(u_{j_k})$ are finite with $\lim_{k \to \infty} \mathcal{F}_0^{(j_k)}(u_{j_k}) = \liminf_{j \to \infty} \mathcal{F}_0^{(j)}(u_j)$,

- $\limsup_{\varepsilon \to 0} \int_\Omega \sup_{|X| \leq M_k} |f_\varepsilon^{(j_k)}(x, X) - f_\varepsilon^{(\infty)}(x, X)| \, dx \leq \frac{1}{k},$

- $\limsup_{\varepsilon \to 0} r_\varepsilon^{(j_k)}(M_k) \leq \frac{1}{k} + \limsup_{j \to \infty} \limsup_{\varepsilon \to 0} r_\varepsilon^{(j)}(M_k) \leq \frac{2}{k}.$

From Lemma (5.4) it follows

$$\infty > \sup_{k \in \mathbb{N}} \mathcal{F}_0^{(j_k)}(u_{j_k}) \geq \alpha \sup_{k \in \mathbb{N}} \|Du_{j_k}\|_{M(\Omega;\mathbb{R}^m)}.$$

Hence, $\{u_{j_k}\}_{k \in \mathbb{N}}$ is bounded in $BV(\Omega;\mathbb{R}^m)$. Then we choose ε_k (with $\varepsilon_k \searrow 0$) so small that there is a $w_k \in L^1(\Omega;\mathbb{R}^m)$ with

- $\|w_k - u_{j_k}\|_{L^1} \leq \frac{1}{j_k},$

- $\mathcal{F}_0^{(j_k)}(u_{j_k}) + \frac{1}{j_k} \geq \mathcal{F}_{\varepsilon_k}^{(j_k)}(w_k),$

- $\int_\Omega \sup_{|X| \leq M_k} |f_{\varepsilon_k}^{(j_k)}(x, X) - f_{\varepsilon_k}^{(\infty)}(x, X)| \, dx \leq \frac{2}{k},$

- $r_{\varepsilon_k}^{(j_k)}(M_k) \leq \frac{3}{k}.$

Notice that $\mathcal{F}_{\varepsilon_k}^{(j_k)}(w_k)$ are finite. Due to our construction, the sequence $\{w_k\}_{k \in \mathbb{N}}$ lies in $W^{1,1}(\Omega;\mathbb{R}^m)$ is there bounded and still converges in $L^1(\Omega;\mathbb{R}^m)$ towards u. We proceed similarly as in Step 1, the only difference being that we now pass in the superscript from j to ∞. Define

$$E_k := \{x \in \Omega : |\nabla w_k| \geq M_k\}.$$

Then

$$
\begin{aligned}
\mathcal{F}_{\varepsilon_k}^{(j_k)}(w_k) &= \int_{\Omega \setminus E_k} f_{\varepsilon_k}^{(j_k)}(x, \nabla w_k(x)) \, dx + \int_{E_k} f_{\varepsilon_k}^{(j_k)}(x, \nabla w_k(x)) \, dx \\
&\geq \int_{\Omega \setminus E_k} f_{\varepsilon_k}^{(\infty)}(x, \nabla w_k(x)) \, dx - \frac{2}{k} + \\
&\quad + \int_{E_k} f_{\varepsilon_k}^{(\infty)}(x, \nabla w_k(x)) \, dx - \frac{3}{k} \int_{E_k} |\nabla w_k(x)| \, dx \\
&\geq \mathcal{F}_{\varepsilon_k}^{(\infty)}(w_k) - \frac{2}{k} - \frac{3}{k} \|\nabla w_k\|_{L^1}.
\end{aligned}
$$

From the construction, the last inequality and the lim inf-inequality for $\{\mathcal{F}_\varepsilon^{(\infty)}\}_\varepsilon$, it follows

$$\liminf_{j \to \infty} \mathcal{F}_0^{(j)}(u_j) = \lim_{k \to \infty} \mathcal{F}_0^{(j_k)}(u_{j_k}) \geq \limsup_{k \to \infty} \mathcal{F}_{\varepsilon_k}^{(j_k)}(w_k) \geq \limsup_{k \to \infty} \mathcal{F}_{\varepsilon_k}^{(\infty)}(w_k) \geq \mathcal{F}_0^{(\infty)}(u).$$

The inequalities from Steps 1 and 2 yield

$$\lim_{j \to \infty} \mathcal{F}_0^{(j)}(u) = \mathcal{F}_0^{(\infty)}(u)$$

for each $u \in L^1(\Omega;\mathbb{R}^m)$.

Step 3: Justification of our assumption.

If we do not assume a priori that $\mathcal{F}_\varepsilon^{(\infty)}$ Γ-converges to $\mathcal{F}_0^{(\infty)}$, by Lemma (5.4), for every subsequence $\{\varepsilon_k\}_{k \in \mathbb{N}}$ there exists a further subsequence $\{\varepsilon_{k_i}\}_{i \in \mathbb{N}}$ such that

$$\Gamma(L^1)\text{-}\lim_{i \to \infty} \mathcal{F}_{\varepsilon_{k_i}}^{(\infty)} =: \mathcal{F}_0^{(\infty)}$$

exists with finite values precisely on $BV(\Omega; \mathbb{R}^m)$. From Steps 1 and 2 it follows that $\mathcal{F}_0^{(\infty)}$ does not depend on the particular subsequence $\{\varepsilon_{k_i}\}_{i \in \mathbb{N}}$. By the Urysohn property for Γ-limits, we thus find that indeed $\Gamma(L^1)\text{-}\lim_{\varepsilon \to 0} \mathcal{F}_\varepsilon^{(\infty)} = \mathcal{F}_0^{(\infty)}$. ∎

Part II

Homogenization in Hencky plasticity with comparison to the small-hardening case

Chapter 6

Hencky plasticity setting

In the second part we will focus on a homogenization problem that is significantly different from the previous. Until now the growth conditions were of standard or at least of Gårding type, always given by some exponent $p \geq 1$ from both sides. Here we will consider densities with different rate of growth in different directions. We will restrict ourselves to the case where the density is bounded from both sides by a multiple of $X \mapsto |X_{\mathrm{dev}}| + (\operatorname{tr} X)^2$. The investigation of such functionals is motivated by their origin in elastoplasticity. Our aim is to loosen the assumptions of the already existing results in this area, and thus to make a first step towards a more general theory. The plan of this chapter is therefore to present very briefly the Hencky plasticity. We will then investigate a related one-dimensional problem and finish by introducing the problem that will be our main focus.

6.1   Hencky plasticity

The case with small strains and stresses was covered by the elasticity theory in the first part. After reaching the so-called yield stress, the material starts to behave plastically. Since by unloading hysteresis effects occur, the problem should be treated as dynamic or at least quasi-static. In the Hencky plasticity it is, however, still considered to be time-independent. Therefore, it must be taken into account that we are modelling a case with one-time loading. The admissible region for the stresses, the elastic region, is given by yield conditions. The common two, the von Mises and Tresca conditions, bound accordingly to experiments the deviatoric part of the stress leaving the hydrostatic part unconstrained.

Thus, the energy of the material has the form

$$\mathcal{E}(u) = \int_{\Omega} \left(\Phi(\mathfrak{E}_{\mathrm{dev}} u(x)) + K(\operatorname{div} u(x))^2 \right) dx$$

where $K > 0$ is a material constant, and $\Phi : \mathbb{R}^{n \times n}_{\mathrm{dev}} \to \mathbb{R}$ is a convex function. We denote for any $X \in \mathbb{R}^{n \times n}$ by

$$X_{\mathrm{dev}} := X - \tfrac{1}{n}(\operatorname{tr} X)I$$

its deviatoric (i.e. zero-trace) part. Accordingly,

$$\mathfrak{E}_{\mathrm{dev}} u := \mathfrak{E} u - \tfrac{1}{n}(\operatorname{div} u)I.$$

To simplify the denotation, $\mathbb{R}^{n\times n}_{\text{dev}}$ will stand for the space of all matrices that are deviatoric and symmetric. The function Φ is quadratic on some compact set and linear on its complement since it is the conjugate of the quadratic form whose domain is the elastic region. Hence, the energy functional has the growth properties described in the introduction.

Let us comment that the structure of the density is very special. It is convex and is a sum of a linearly growing and a quadratic term. We wish to explore a more general situation without the splitting and (possibly) without convexity. Moreover, we are interested in homogenization problems for fine mixtures so we allow for location-dependent periodic densities.

Being very short, we refer to the extensive literature for a more precise description. For a survey on plasticity theory, consult [HR:99] and Chapter V of [DL:76]. The relaxation of the basic problem described above is treated in Section II.6 of [Te:85] and in the articles [AG:82] and [BDV:97]. The location-dependent case with a general convex function having the described growth was covered in [DQ:90]. We will present this result extensively later in Subsection 8.4.2.

The setting that we have just described corresponds to the model of perfect plasticity. There one supposes that the stress-strain curve consists of a linear (i.e. elastic) and a constant part. This means that after reaching the yield stress the deformation grows at constant stress. Such behaviour is also known as zero hardening. See, e.g., Figure 3.2 in [HR:99].

We would like to consider also the setting with hardening. The simplest model is to have also in the plastic regime a linear function. If the slope is very flat, we speak of a small hardening. In this case, the energy has a small quadratic growth also in the deviatoric part, which changes the setting essentially. We will be interested in the question if the zero-hardening model may be regarded as a limit of the model with hardening when sending the slope of the plastic part to 0. See also Section 3.5 in [HN:80]. Thus, we again have two processes: the homogenization and the vanishing hardening.

6.2   One-dimensional toy model

Let us first consider a very simple model. We will have a one-dimensional setting with a composite of two materials. We suppose that their stress-strain diagram first has the elastic regime, determined by the Young modulus, and then the plastic regime with or without hardening. For the sake of simplicity, we suppose the hardening inclination to be the same for both materials. Therefore, let

$$V_i^{(\delta)}(\xi) = \begin{cases} \frac{1}{2}\alpha_i\xi^2, & |\xi| < \xi_i, \\ \frac{1}{2}\delta\xi^2 + (\alpha_i - \delta)\xi_i|\xi| - \frac{1}{2}(\alpha_i - \delta)\xi_i^2, & |\xi| > \xi_i, \end{cases}$$

for $i = 1, 2$ represent the energy density for each material. Here $\alpha_i > 0$ denotes the corresponding Young modulus and $\xi_i > 0$ gives the yield point, while $\delta \geq 0$ determines the rate of hardening. The setting becomes even more apparent by noticing

$$(V_i^{(\delta)})''(\xi) = \begin{cases} \alpha_i, & |\xi| < \xi_i, \\ \delta, & |\xi| > \xi_i. \end{cases}$$

Of course, this problem is not a special case of the one from the previous section since the trace part in dimension one represents the entire "matrix". However, we would first like to understand this simple setting and explicitly determine the homogenized density.

Let the first material occupy the portion of the mixture given by $0 < a < 1$. Thus, the density $V^{(\delta)} : \mathbb{R} \times \mathbb{R} \to \mathbb{R}$ is 1-periodic function with values

$$V^{(\delta)}(x, \xi) = \begin{cases} V_1^{(\delta)}(\xi), & 0 < x < a, \\ V_2^{(\delta)}(\xi), & a < x < 1, \end{cases}$$

for $x \in \mathbb{I}$. Let $J \subset \mathbb{R}$ be an open bounded interval. We wish to determine the behaviour of the integral functional

$$\mathcal{F}_\varepsilon^{(\delta)}(u) := \int_J V^{(\delta)}(\tfrac{x}{\varepsilon}, u'(x)) \, dx.$$

as $\varepsilon \to 0$. For $\delta > 0$ its natural domain is $W^{1,2}(J)$, whereas for $p = 1$ we may even take $W^{1,1}(J)$. Notice that the density is convex in the second variable, and therefore the one-cell formula may be employed.

6.2.1 Convex homogenization

We start by making a more general observation. Let $J \subset \mathbb{R}$ be a bounded open interval and $p \geq 1$. Take convex functions $f_i : \mathbb{R} \to \mathbb{R}$, $i = 1, 2$, that fulfil the standard p-growth condition, and define for any $0 < a < 1$ a 1-periodic function $f : \mathbb{R} \times \mathbb{R} \to \mathbb{R}$ by setting for $x \in \mathbb{I}$

$$f(x, \xi) = \begin{cases} f_1(\xi), & x < a, \\ f_2(\xi), & x > a, \end{cases}$$

and extending it periodically to the whole $\mathbb{R}$. Then the functionals $\mathcal{F}_\varepsilon : L^p(J) \to \mathbb{R} \cup \{\infty\}$ where

$$\mathcal{F}_\varepsilon(u) := \begin{cases} \int_J f(\tfrac{x}{\varepsilon}, u'(x)) \, dx, & u \in W^{1,p}(J), \\ \infty, & \text{else}, \end{cases}$$

$\Gamma(L^p)$-converge to the functional $\mathcal{F}_{\mathrm{hom}}$ given for $p > 1$ by

$$\mathcal{F}_{\mathrm{hom}}(u) := \begin{cases} \int_J f_{\mathrm{hom}}(u'(x)) \, dx, & u \in W^{1,p}(J), \\ \infty, & \text{else}. \end{cases}$$

Having a scalar case with a convex density, this was proved already in [Ma:78]. The exception is the case $p = 1$ with domain being $BV(J)$, and for $u \in BV(J) \setminus W^{1,1}(J)$ we get

$$\mathcal{F}_{\mathrm{hom}}(u) = \int_J f_{\mathrm{hom}}(u'(x)) \, dx + \int_J (f_{\mathrm{hom}})^\infty \big(\tfrac{dD^s u}{d|D^s u|}(x)\big) d|D^s u|$$

(see, e.g., Theorem 12.3.2 in [ABM:06]). The recession function is as in Chapter 5. In all cases the new density is given by the one-cell formula

$$f_{\mathrm{hom}}(\xi) = \inf_{\varphi \in W_0^{1,p}(0,1)} \int_0^1 f(x, \xi + \varphi'(x)) \, dx.$$

Therefore,

$$
\begin{aligned}
f_{\hom}(\xi) &= \inf_{\varphi \in W_0^{1,p}(0,1)} \int_0^1 f(x, \xi + \varphi'(x))\, dx \\
&= \inf_{\varphi \in W_0^{1,p}(0,1)} \left(\int_0^a f_1(\xi + \varphi'(x))\, dx + \int_a^1 f_2(\xi + \varphi'(x))\, dx \right) \\
&= \inf_{t \in \mathbb{R}} \inf \left\{ \int_0^a f_1(\xi + \varphi'(x))\, dx + \int_a^1 f_2(\xi + \varphi'(x))\, dx : \varphi \in W_0^{1,p}(0,1), \varphi(a) = t \right\} \\
&= \inf_{t \in \mathbb{R}} \left(\inf_{\varphi \in W_0^{1,p}(0,a)} \int_0^a f_1(\xi + \tfrac{t}{a} + \varphi'(x))\, dx + \inf_{\varphi \in W_0^{1,p}(a,1)} \int_a^1 f_2(\xi - \tfrac{t}{1-a} + \varphi'(x))\, dx \right) \\
&= \inf_{t \in \mathbb{R}} \left(a f_1(\xi + \tfrac{t}{a}) + (1-a) f_2(\xi - \tfrac{t}{1-a}) \right).
\end{aligned}
$$

Define

$$
M_\xi(t) := a f_1(\xi + \tfrac{t}{a}) + (1-a) f_2(\xi - \tfrac{t}{1-a}).
$$

Then

$$
\partial M_\xi(t) = \partial f_1(\xi + \tfrac{t}{a}) - \partial f_2(\xi - \tfrac{t}{1-a}).
$$

Let t_ξ minimize the expression in the definition of $f_{\hom}(\xi)$. Hence, it is determined by

$$
\partial f_1(\xi + \tfrac{t_\xi}{a}) \cap \partial f_2(\xi - \tfrac{t_\xi}{1-a}) \neq \emptyset,
$$

or if we have differentiable functions by

$$
f_1'(\xi + \tfrac{t_\xi}{a}) = f_2'(\xi - \tfrac{t_\xi}{1-a}).
$$

6.2.2   Zero hardening

Now we return to our setting. First we explore the zero-hardening case, when $\delta = 0$. (We omit writing the superscript $^{(0)}$.) Let us henceforth suppose

$$
\alpha_2 \xi_2 > \alpha_1 \xi_1.
$$

In the picture we have the possible graphs of V_i' or, mechanically, the stress-strain diagrams for both materials:

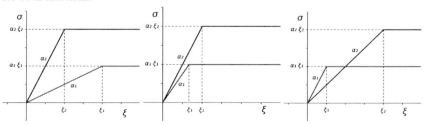

The analysis is, however, the same for all three cases.

According to the previous subsection, we must to every $\xi > 0$ find t_ξ such that

$$V_1'(\xi + \tfrac{t_\xi}{a}) = V_2'(\xi - \tfrac{t_\xi}{1-a}).$$

On the interval $(\frac{\alpha_1 \xi_1}{\alpha_2}, \infty)$ the function V_2' is larger than $\alpha_1 \xi_1 = \max V_1'$. Therefore, t_ξ fulfils $\xi - \frac{t_\xi}{1-a} \leq \frac{\alpha_1 \xi_1}{\alpha_2}$, and consequently

$$V_2(\xi - \tfrac{t_\xi}{1-a}) = \tfrac{\alpha_2}{2}(\xi - \tfrac{t_\xi}{1-a})^2.$$

So, with V_2 we always land in the quadratic regime. For V_1 both behaviours can occur. Since the ratio of the shifts in V_1' and V_2' is fixed, i.e. $(1-a) : a$, there exists a switch point ξ_k that determines the regime of V_1 (see also the pictures): for $\xi < \xi_k$ also $V_1(\xi + \tfrac{t_\xi}{a})$ is still quadratic, whereas for $\xi > \xi_k$ it is already linear. Therefore, ξ_k is given by the linear system

$$\xi_k - \frac{t_{\xi_k}}{1-a} = \frac{\alpha_1 \xi_1}{\alpha_2},$$

$$\xi_k + \frac{t_{\xi_k}}{a} = \xi_1.$$

It follows

$$\xi_k = (1-a)\frac{\alpha_1 \xi_1}{\alpha_2} + a\xi_1 = \alpha_1 \xi_1 \left(\frac{1-a}{\alpha_2} + \frac{a}{\alpha_1}\right) = \frac{\alpha_1 \xi_1}{\overline{\alpha}}$$

where

$$\overline{\alpha} := \left(\frac{a}{\alpha_1} + \frac{1-a}{\alpha_2}\right)^{-1}$$

is the weighted harmonic mean of α_1 and α_2. Now let us look into two regimes:

- $\xi < \xi_k$:

   In this case the situation is the same as if we had two parabolas. The result is known (see, e.g., Subsection 12.3.1 in [ABM:06]). We get a part of a parabola with the leading coefficient given by the weighted harmonic mean. For the sake of completeness, we carry on with computation. We have

   $$\alpha_1(\xi + \tfrac{t_\xi}{a}) = V_1'(\xi + \tfrac{t_\xi}{a}) = V_2'(\xi - \tfrac{t_\xi}{1-a}) = \alpha_2(\xi - \tfrac{t_\xi}{1-a}).$$

   Hence,

   $$t_\xi = \frac{(\alpha_2 - \alpha_1)\xi}{\frac{\alpha_1}{a} + \frac{\alpha_2}{1-a}},$$

   and

   $$\alpha_1(\xi + \tfrac{t_\xi}{a}) = \overline{\alpha}\xi = \alpha_2(\xi - \tfrac{t_\xi}{1-a}).$$

   It follows

   $$\begin{aligned} V_{\text{hom}}(\xi) &= M_\xi(t_\xi) \\ &= a\tfrac{\alpha_1}{2}(\xi + \tfrac{t_\xi}{a})^2 + (1-a)\tfrac{\alpha_2}{2}(\xi - \tfrac{t_\xi}{1-a})^2 \\ &= \tfrac{1}{2}\left(\tfrac{a}{\alpha_1} + \tfrac{1-a}{\alpha_2}\right)(\overline{\alpha}\xi)^2 \\ &= \tfrac{1}{2}\overline{\alpha}\xi^2. \end{aligned}$$

- $\xi > \xi_k$:

  There
  $$\alpha_1\xi_1 = V_1'(\xi + \tfrac{t_\xi}{a}) = V_2'(\xi - \tfrac{t_\xi}{1-a}) = \alpha_2(\xi - \tfrac{t_\xi}{1-a}).$$

  It follows
  $$\xi - \tfrac{t_\xi}{1-a} = \tfrac{\alpha_1\xi_1}{\alpha_2},$$

  and
  $$\xi + \tfrac{t_\xi}{a} = \tfrac{1}{a}\Big(\xi - (1-a)\tfrac{\alpha_1\xi_1}{\alpha_2}\Big).$$

  Therefore,
  $$
  \begin{aligned}
  V_{\text{hom}}(\xi) &= M_\xi(t_\xi)\\
  &= a\alpha_1\xi_1(\xi + \tfrac{t_\xi}{a} - \tfrac{\xi_1}{2}) + (1-a)\tfrac{\alpha_2}{2}(\xi - \tfrac{t_\xi}{1-a})^2\\
  &= a\alpha_1\xi_1\Big(\tfrac{1}{a}(\xi - (1-a)\tfrac{\alpha_1\xi_1}{\alpha_2}) - \tfrac{\xi_1}{2}\Big) + (1-a)\tfrac{\alpha_2}{2}(\tfrac{\alpha_1\xi_1}{\alpha_2})^2\\
  &= \alpha_1\xi_1\xi + \alpha_1\xi_1^2\Big(-(1-a)\tfrac{\alpha_1}{\alpha_2} - \tfrac{a}{2} + (1-a)\tfrac{\alpha_1}{2\alpha_2}\Big)\\
  &= \alpha_1\xi_1\xi - \tfrac{1}{2}\alpha_1^2\xi_1^2\tfrac{1}{\overline{\alpha}}\\
  &= \alpha_1\xi_1\Big(\xi - \tfrac{1}{2}\tfrac{\alpha_1\xi_1}{\overline{\alpha}}\Big).
  \end{aligned}
  $$

Altogether we have shown that
$$
V_{\text{hom}}(\xi) = \left\{
\begin{array}{ll}
\tfrac{1}{2}\overline{\alpha}\xi^2, & |\xi| < \tfrac{\alpha_1\xi_1}{\overline{\alpha}},\\
\alpha_1\xi_1(|\xi| - \tfrac{1}{2}\tfrac{\alpha_1\xi_1}{\overline{\alpha}}), & |\xi| > \tfrac{\alpha_1\xi_1}{\overline{\alpha}}.
\end{array}
\right.
$$

The homogenized function still possesses two regimes: the quadratic and the linear. The leading coefficient of the parabola is the weighted harmonic mean of the original coefficients, and the switch point lies at ξ_k.

With our assumption $\alpha_2\xi_2 > \alpha_1\xi_1$ we excluded the case when the products are equal. Checking all the computations, we see that the result holds also in that case. The formula for ξ_k can be in that case rewritten as $\xi_k = a\xi_1 + (1-a)\xi_2$.

Let us conclude this discussion by gathering the results.

Lemma (6.1). *Let $0 < a < 1$ and $\alpha_1, \alpha_2, \xi_1, \xi_2 > 0$ be arbitrary with $\alpha_2\xi_2 \geq \alpha_1\xi_1$. Consider the function $V : \mathbb{R} \times \mathbb{R} \to \mathbb{R}$ that is 1-periodic in the first argument and is for $x \in \mathbb{I}$ given by*
$$
V(x,\xi) = \left\{
\begin{array}{ll}
V_1(\xi), & x < a,\\
V_2(\xi), & x > a,
\end{array}
\right.
$$

with $V_i = \alpha_i\Big(\tfrac{|\cdot|^2}{2} \wedge \xi_i|\cdot|\Big)^{**}$. Its homogenization is
$$
V_{\text{hom}} = \overline{\alpha}\Big(\tfrac{|\cdot|^2}{2} \wedge \xi_k|\cdot|\Big)^{**}
$$

where
$$
\tfrac{1}{\overline{\alpha}} = \tfrac{a}{\alpha_1} + \tfrac{1-a}{\alpha_2} \quad and \quad \xi_k = \tfrac{\alpha_1\xi_1}{\overline{\alpha}}.
$$

We used the customary denotations: f^{**} for the convex envelope of f and $a \wedge b = \min\{a, b\}$.

This result complements the purely quadratic one-dimensional case presented, e.g., in Subsection 12.3.1 of [ABM:06]). As already mentioned, in the case of two parabolas, the result is again a parabola, and its leading coefficient is the weighted harmonic mean of the corresponding leading coefficients.

6.2.3   Small hardening

Now consider a small $\delta > 0$. We may also write

$$V_i^{(\delta)}(\xi) = \tfrac{1}{2}(\alpha_i \xi^2 - (\alpha_i - \delta)(|\xi| - \xi_i)_+^2)$$

where $\xi_+ = \max\{\xi, 0\}$. The key role play the derivatives

$$(V_i^{(\delta)})'(\xi) = \left\{ \begin{array}{ll} \alpha_i \xi, & |\xi| < \xi_i, \\ \alpha_i \xi_i + \delta(\xi - \xi_i), & |\xi| > \xi_i. \end{array} \right.$$

There are three different regimes. The first switch point is (as before) at

$$\xi_k = \frac{\alpha_1 \xi_1}{\bar{\alpha}}.$$

The case $0 < \xi < \xi_k$ remains exactly the same as when $\delta = 0$. Then the interval follows where $\xi + \frac{t_\xi}{a} > \xi_1$ (already the plastic regime), whereas $\xi - \frac{t_\xi}{1-a} < \xi_2$ (still the elastic regime). This ends at the second switch point $\xi_l^{(\delta)}$ determined by

$$\begin{aligned} \xi_l^{(\delta)} - \frac{t_{\xi_l^{(\delta)}}}{1 - a} &= \xi_2, \\ \xi_l^{(\delta)} + \frac{t_{\xi_l^{(\delta)}}}{a} &= \xi_1 + \frac{\alpha_2 \xi_2 - \alpha_1 \xi_1}{\delta}. \end{aligned}$$

Hence

$$\xi_l^{(\delta)} = a\xi_1 + (1 - a)\xi_2 + a\frac{\alpha_2 \xi_2 - \alpha_1 \xi_1}{\delta}.$$

For $\xi > \xi_l^{(\delta)}$ both minimizing expressions lie in the corresponding plastic regime. It is inconvenient to give the explicit formula for V_{hom}. It is a C^1-function consisting of three parabolas with leading coefficients given by

$$(V_{\text{hom}}^{(\delta)})''(\xi) = \left\{ \begin{array}{ll} (\frac{a}{\alpha_1} + \frac{1-a}{\alpha_2})^{-1}, & |\xi| < \xi_k, \\ (\frac{a}{\delta} + \frac{1-a}{\alpha_2})^{-1}, & \xi_k < |\xi| < \xi_l^{(\delta)}, \\ \delta, & |\xi| > \xi_l^{(\delta)}. \end{array} \right.$$

Hence, first we have the weighted harmonic mean (in this order) of α_1 and α_2, then of δ and α_2, and finally of δ and δ. Of course, by inserting $\delta = 0$ we recover the results for zero hardening.

6.2.4   Commutability of homogenization and vanishing hardening

Having determined the homogenized densities, we may now look into the problem of interchangeability of homogenization and vanishing hardening for this very special case. First, let us fully define the integral functionals in question. Although the cases with small and zero hardening differ significantly, we should choose the same ambient space (and the same domain) in order to compare them. In view of the zero-hardening case, we take as the ambient space $L^1(J)$, and we set $W^{1,2}(J)$ as domain. For $u \in W^{1,2}(J)$ let

$$\mathcal{F}_\varepsilon^{(\delta)}(u) := \int_J V^{(\delta)}(\tfrac{x}{\varepsilon}, u'(x)) \, dx.$$

Schematically, we again investigate the diagrams of form

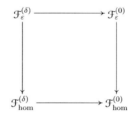

with the lower row still to be determined.

Clearly we have for $\delta > 0$

$$\Gamma(L^2)\text{-}\lim_{\varepsilon \to 0} \mathcal{F}_\varepsilon^{(\delta)} = \mathcal{F}_{\text{hom}}^{(\delta)}$$

where for $u \in W^{1,2}(J)$

$$\mathcal{F}_{\text{hom}}^{(\delta)}(u) := \int_J V_{\text{hom}}^{(\delta)}(u'(x)) \, dx.$$

By enlarging the ambient space to $L^1(J)$, the domain remains the same due to the quadratic lower bound. Since for $u \in W^{1,2}(J)$ the Poincaré inequality

$$\|u\|_{L^2} \leq C(\|u'\|_{L^2} + \|u\|_{L^1})$$

holds, we conclude that actually

$$\Gamma(L^1)\text{-}\lim_{\varepsilon \to 0} \mathcal{F}_\varepsilon^{(\delta)} = \mathcal{F}_{\text{hom}}^{(\delta)}.$$

For the upper row: since $V^{(\delta)} \searrow V$ pointwisely, by the dominated convergence theorem also

$$\mathcal{F}_\varepsilon^{(\delta)} \searrow \mathcal{F}_\varepsilon^{(0)},$$

or in terms of Γ-convergence

$$\Gamma(L^1)\text{-}\lim_{\delta \to 0} \mathcal{F}_\varepsilon^{(\delta)} = \text{lsc } \mathcal{F}_\varepsilon^{(0)}.$$

When relaxing, we should be aware of the fact that the density is location-dependent and is not continuous in x. Such a setting is covered in Theorem 4.1.4 of [BFM:98]) with explicit formulas. The regular part is in our case still determined by the density V, and the singular part may be computed from the recession function

$$V^\infty(x,\xi) = \lim_{t\to\infty} \tfrac{1}{t} V(x, t\xi) = \begin{cases} \alpha_1 \xi_1 |\xi|, & x < a, \\ \alpha_2 \xi_2 |\xi|, & x > a. \end{cases}$$

Currently, we have

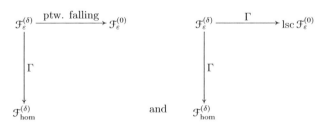

Now consider the Γ-convergence of $\mathcal{F}_\varepsilon^{(0)}$ or $\text{lsc}\,\mathcal{F}_\varepsilon^{(0)}$. The Γ-limit is given by

$$\mathcal{F}_{\text{hom}}^{(0)}(u) = \int_J V_{\text{hom}}(u'(x))\,dx + \int_J (V_{\text{hom}})^\infty (\tfrac{dD^s u}{d|D^s u|}(x))d|D^s u|(x)$$

for $u \in BV(J)$. By a density argument, it makes no difference that the domain of $\mathcal{F}_\varepsilon^{(0)}$ is $W^{1,2}$ instead of $W^{1,1}$. Again we may compute

$$(V_{\text{hom}})^\infty(\xi) = \lim_{t\to\infty} \tfrac{1}{t} V_{\text{hom}}(t\xi) = \alpha_1 \xi_1 |\xi|.$$

Hence,

$$\mathcal{F}_{\text{hom}}^{(0)}(u) = \int_J \frac{\bar\alpha}{2}\Big(u'(x)^2 - (|u'(x)| - \xi_k)_+^2\Big)\,dx + \alpha_1 \xi_1 |D^s u|(J).$$

So far, we have got

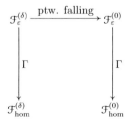

Since $(V_{\text{hom}}^{(\delta)})''$ descends pointwisely to V_{hom}'', also $V_{\text{hom}}^{(\delta)}$ descends pointwisely to V_{hom}. Therefore, the same holds for the corresponding functionals (with domain $W^{1,2}(\Omega)$). For Γ-convergence, we have to relax the functional as in the first row above which again yields

$\mathcal{F}_{\mathrm{hom}}^{(0)}$. Therefore

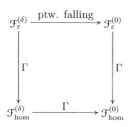

6.3 Setting the problem

Our aim in the following will be to consider the setting typical for the Hencky plasticity and to make an analogous analysis as for the one-dimensional case above. We will generalize the assumptions to allow for more general applications: The ($\mathbb{I}^n$-periodic) density for the zero hardening $V : \mathbb{R} \times \mathbb{R}_{\mathrm{sym}}^{n \times n} \to \mathbb{R}$ will neither split in two parts nor be everywhere convex. We will only assume that it has a growth typical for the Hencky plasticity, i.e., that there exist $\alpha, \beta > 0$ such that

$$\alpha(|X_{\mathrm{dev}}| + (\operatorname{tr} X)^2) \leq V(x, X) \leq \beta(|X_{\mathrm{dev}}| + (\operatorname{tr} X)^2 + 1)$$

for a.e. $x \in \mathbb{R}^n$ and all $X \in \mathbb{R}_{\mathrm{sym}}^{n \times n}$. We will refer to this property as a *Hencky plasticity growth*. However, we will later have to demand some kind of convexity at infinity. We will model the small hardening by adding a quadratic term, i.e.,

$$V^{(\delta)}(x, X) := V(x, X) + \delta |X_{\mathrm{dev}}|^2.$$

Let us define

$$\mathcal{F}_{\varepsilon}^{(\delta)}(u) := \int_{\Omega} V^{(\delta)}(\tfrac{x}{\varepsilon}, \mathfrak{E}u(x)) \; dx$$

for $u \in W^{1,2}(\Omega; \mathbb{R}^n)$. Again we notice that a larger domain can be taken for $\delta = 0$. The appropriate space, i.e. $LU(\Omega; \mathbb{R}^n)$, will be defined in Chapter 7 as well as its weak closure.

We will investigate the behaviour of the functionals when we send either of the two small parameters to 0: By sending $\delta \to 0$ we are dealing with vanishing hardening, whereas $\varepsilon \to 0$ corresponds to the effective functionals for fine mixtures. As before, we would also like to find out whether these processes are interchangeable.

Outlining the discussion schematically, we wish to explore a potential commutability of the corresponding diagram

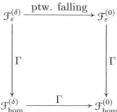

The right ambient space is $L^1(\Omega; \mathbb{R}^n)$. For the left arrow downwards, the situation is very much clear. The case was considered in Chapter 2, as we are dealing with functionals of Gårding type. Additionally, a similar step from L^2 to L^1 as in the one-dimensional case must be done. The bottom row inherits monotonicity from the upper, which reduces the determination of Γ-convergence to relaxation.

Hence, the plan is as follows. As already announced, we must look into the structure of functions that form domains of the functionals in the right column. Moreover, we expect some kind of homogenization process also for them, but we have yet to define the appropriate density as the functionals are not even of Gårding type. This will be the most important and difficult issue: to understand the homogenizing process for function with a Hencky plasticity growth. In that sense, adding a small hardening may mathematically also be regarded as a regularization of the zero-hardening setting. Namely, thus we get functionals with better growth conditions.

Let us at the end remark that everything is still valid if we model the hardening by any Carathéodory functions $V^{(\delta)} : \mathbb{R}^n \times \mathbb{R}^{n \times n}_{\text{sym}} \to \mathbb{R}$ with quadratic lower and upper bound such that

$$V^{(\delta)} \searrow V \quad \text{and} \quad V^{(\delta)}(x, X) - V(x, X) \leq C\delta |X|^2.$$

Chapter 7

Functions of bounded deformation

In the preceding chapter we presented the setting. Before continuing, we must define and explore spaces of functions that are the natural domains of the integral functionals in question. We notice that only the symmetrized gradient must be well-defined and be integrable with the right exponent. For exponents greater than 1, this does not lead to larger spaces since Korn's inequality holds, and we get the corresponding Sobolev space as domain. For $p = 1$, however, we will introduce the space LD, which is larger than the matching Sobolev space $W^{1,1}$. They share the same deficit though, i.e., boundedness does not imply weak compactness. This leads to the definition of functions of bounded deformation, which correspond to functions of bounded variation for the full-gradient case. We will present some basic facts regarding convergence, approximation, boundary values and embeddings. Two results will be of great importance later: the analogue of Alberti's rank one theorem about the structure of the singular part of the symmetrized gradient and approximate differentiability. We will conclude the first section by determining functions with zero deviatoric part of the symmetrized gradient as to give an alternative form of the Poincaré-Korn inequality.

In the Hencky plasticity setting, the density has a quadratic growth in the trace part. Therefore, we must narrow the space BD by imposing the divergence to be square-integrable. That leads to the introduction of spaces LU and U. (Note that the denotation varies in the literature.) We will present the relevant results also for these spaces in Section 7.2 and show a version of the Helmholtz decomposition to hold. At the end of the chapter, we will discuss existence of functions with zero boundary values and prescribed divergence.

This chapter serves as a preparation for our analysis, and so it mostly consists of known results with citations. Therefore, let us clearly indicate our own contributions. We include our proof of Proposition (7.7) solely for its simplicity. A higher order approximate differentiability will be needed essentially in Chapter 8. In the recent paper [ABC:14], L^q-differentiability of BD-functions was shown for $1 \leq q < \frac{n}{n-1}$. The authors explicitly state that their work does not apply to the boundary case $q = \frac{n}{n-1}$. For the sake of completeness, we therefore include the proof of $L^{\frac{n}{n-1}}$-differentiability since it can be derived fairly simply from L^1-differentiability, proved already in [ACD:97], adapting the strategy for the same step for BV-functions in [AFP:00]. The Helmholtz decomposition in Subsection 7.2.2 is, to the best of our knowledge, new and shows an interesting structure of the space U.

7.1   Functions of bounded deformation

7.1.1   Definition and basic properties

We may define $\mathfrak{E}u$ via distribution or directly as the only (if existing) function that fulfils

$$\int_\Omega \mathfrak{E}u(x) \cdot \Phi(x)\ dx = -\int_\Omega u(x) \cdot \operatorname{div}(\Phi(x)_{\mathrm{sym}})\ dx$$

for all $\Phi \in C_c^\infty(\Omega; \mathbb{R}^{n\times n})$. We will denote the corresponding space by

$$LD(\Omega; \mathbb{R}^n) := \{u \in L^1(\Omega; \mathbb{R}^n) : \mathfrak{E}u \in L^1(\Omega; \mathbb{R}^{n\times n})\}.$$

It becomes a Banach space when equipped with the natural norm

$$\|u\|_{LD(\Omega;\mathbb{R}^n)} := \|u\|_{L^1(\Omega;\mathbb{R}^n)} + \|\mathfrak{E}u\|_{L^1(\Omega;\mathbb{R}^{n\times n})}.$$

For the properties of this space, we refer to, e.g., Section II.1 in [Te:85]. Analogously to the space of functions of bounded variation for the full-gradient case, one introduces the space of functions of bounded deformation as follows: If the mapping

$$C_c^\infty(\Omega; \mathbb{R}^{n\times n}) \to \mathbb{R}, \quad \Phi \mapsto -\int_\Omega u(x) \cdot \operatorname{div}(\Phi(x)_{\mathrm{sym}})\ dx,$$

may be extended to a bounded linear functional on $C_0(\Omega; \mathbb{R}^{n\times n})$, i.e. to a Radon measure, then we denote this functional by $Eu \in M(\Omega; \mathbb{R}^{n\times n})$. (Clearly, it must lie in $M(\Omega; \mathbb{R}^{n\times n}_{\mathrm{sym}})$.) The space of functions of bounded deformation is defined by

$$BD(\Omega; \mathbb{R}^n) := \{u \in L^1(\Omega; \mathbb{R}^n) : Eu \in M(\Omega; \mathbb{R}^{n\times n})\}.$$

Equipped with the norm

$$\|u\|_{BD(\Omega;\mathbb{R}^n)} := \|u\|_{L^1(\Omega;\mathbb{R}^n)} + \|Eu\|_{M(\Omega;\mathbb{R}^{n\times n})},$$

it is also a Banach space.

As in the space of functions of bounded variation, we introduce (with abuse of terminology) the weak convergence: A sequence $\{u_j\}_{j\in\mathbb{N}}$ *converges weakly in $BD(\Omega;\mathbb{R}^n)$* to u, denoted $u_j \rightharpoonup u$, if

$$u_j \to u \quad \text{in } L^1(\Omega;\mathbb{R}^n) \quad \text{and} \quad Eu_j \overset{*}{\rightharpoonup} Eu \quad \text{in } M(\Omega;\mathbb{R}^{n\times n}).$$

Let us mention that every bounded sequence in $BD(\Omega;\mathbb{R}^n)$ contains a weakly convergent subsequence.

If additionally $|Eu_j|(\Omega) \to |Eu|(\Omega)$, then we speak of *strict* or *intermediate* convergence. This topology is actually induced by the metric

$$d(u,v) := \|u - v\|_{L^1} + \big||Eu|(\Omega) - |Ev|(\Omega)\big|.$$

Finally, let $c : \mathbb{R}^{n\times n}_{\mathrm{sym}} \to [0, \infty)$ be a convex function with linear upper bound. A sequence $\{u_j\}_{j\in\mathbb{N}}$ *converges c-strictly in $BD(\Omega;\mathbb{R}^n)$* to u, symbolically $u_j \overset{c}{\rightharpoonup} u$, if

- $u_j \to u$ in $L^1(\Omega;\mathbb{R}^n)$,

- $|Eu_j|(\Omega) \to |Eu|(\Omega)$,

- $\int_\Omega c(Eu_j) \to \int_\Omega c(Eu)$.

For a general discussion about a convex linearly bounded function of a measure including the density results for c-strict topology, see Section II.4 in [Te:85] or [DT:84]. Let us just mention that every $\mu \in M(\Omega; \mathbb{R}^{n \times n}_{\mathrm{sym}})$ has the Lebesgue decomposition

$$\mu = \mu^a + \mu^s$$

into the absolutely continuous and the singular part with respect to the Lebesgue measure $\mathcal{L}^n$. Then

$$c(\mu) := c\Big(\frac{d\mu^a}{d\mathcal{L}^n}\Big)\mathcal{L}^n + c^\infty\Big(\frac{d\mu^s}{d|\mu^s|}\Big)|\mu^s|.$$

Moreover, the denotation $\int_A c(\mu)$ used above is merely another way for writing $c(\mu)(A)$. For $u \in BD(\Omega; \mathbb{R}^n)$ we decompose Eu with respect to the Lebesgue measure $\mathcal{L}^n$ into

$$Eu =: \mathfrak{E}u \, \mathcal{L}^n + E^s u.$$

By denoting the density of the absolutely continuous part by $\mathfrak{E}u$, we have just extended the definition from $LD(\Omega; \mathbb{R}^n)$. Clearly,

$$LD(\Omega; \mathbb{R}^n) = \{u \in BD(\Omega; \mathbb{R}^n) : E^s u = 0\}.$$

Although being larger than the space of functions of bounded variation, some properties still hold also for the space of functions of bounded deformation. E.g., it is possible to define boundary values, as was shown in Theorem II.2.1 of [Te:85] for domains with C^1-boundary and extended recently by Babadjian, see [Ba:15]:

Theorem (7.1). *Let $\Omega \subset \mathbb{R}^n$ be a bounded Lipschitz domain. There exists a unique linear continuous mapping $\gamma : BD(\Omega; \mathbb{R}^n) \to L^1(\partial\Omega; \mathbb{R}^n)$ such that the following integration by parts formula holds: for every $u \in BD(\Omega; \mathbb{R}^n)$ and $\varphi \in C^1(\mathbb{R}^n)$*

$$\int_\Omega u(x) \odot \nabla\varphi(x) \, dx + \int_\Omega \varphi(x) \, dEu(x) = \int_{\partial\Omega} \varphi(x) \, \gamma(u)(x) \odot \nu(x) \, d\mathscr{H}^{n-1}(x).$$

For all $u \in BD(\Omega; \mathbb{R}^n) \cap C(\overline{\Omega}; \mathbb{R}^n)$, it holds $\gamma(u) = u|_{\partial\Omega}$. In point of fact, γ is even continuous if $BD(\Omega; \mathbb{R}^n)$ is endowed with strict topology.

Here ν denotes the outer unit normal vector to $\partial\Omega$ and $\mathscr{H}^{n-1}$ the $(n-1)$-dimensional Hausdorff measure. Henceforth, we will for $a, b \in \mathbb{R}^n$ write

$$a \odot b := \tfrac{1}{2}(a \otimes b + b \otimes a).$$

Theorems II.2.2 and II.2.4 in [Te:85] cover the subject of embeddings in L^p-spaces, see also [AG:80]:

Theorem (7.2). *Let $\Omega \subset \mathbb{R}^n$ be a bounded Lipschitz domain. There exists a natural bounded embedding*

$$BD(\Omega; \mathbb{R}^n) \hookrightarrow L^{\frac{n}{n-1}}(\Omega; \mathbb{R}^n).$$

For every $1 \le q < \frac{n}{n-1}$ the embedding

$$BD(\Omega; \mathbb{R}^n) \hookrightarrow L^q(\Omega; \mathbb{R}^n)$$

is even compact.

7.1.2   Alberti's rank one theorem

The singular part of the gradient of a function with bounded variation has a rank-one structure. This was proved in [Al:93] and is commonly known as Alberti's rank one theorem. It has been used in blow-up techniques for a whole range of variational problems in the corresponding space. For a long period it was unknown whether functions of bounded deformation possess the analogue property until De Philippis and Rindler showed the conjecture to hold (see Theorem 1.7 in [DR:16]):

Theorem (7.3). *Let $\Omega \subset \mathbb{R}^n$ be an open set and let $u \in BD(\Omega; \mathbb{R}^n)$. Then, for $|E^s u|$-a.e. $x \in \Omega$, there exist $a(x), b(x) \in \mathbb{R}^n \setminus \{0\}$ such that*

$$\frac{dE^s u}{d|E^s u|} = a(x) \odot b(x).$$

7.1.3   Approximate differentiability

Although for the functions of bounded deformation the (full) gradient is in general not even a Radon measure, they still can be locally (on average) approximated by linear functions. More precisely,

Theorem (7.4). *For every $u \in BD(\Omega; \mathbb{R}^n)$ there exists a negligible set $N \subset \Omega$ such that for all $x_0 \in \Omega \setminus N$ there exists a matrix $L_{x_0} \in \mathbb{R}^{n \times n}$ such that*

$$\lim_{r \to 0} \frac{1}{r^n} \int_{B_r(x_0)} \frac{|u(x) - u(x_0) - L_{x_0}(x - x_0)|}{r} \, dx = 0.$$

Therefore, u is a.e. approximately differentiable with $L_{x_0} = \nabla u(x_0)$ being the approximate differential. Moreover, it holds $\mathfrak{E}u(x) = \mathrm{sym}\,\nabla u(x)$ for a.e. $x \in \Omega$. The function ∇u is in the weak-L^1-space since

$$|\{x : |\nabla u(x)| > t\}| \leq \frac{c(n)}{t} |Eu|(\Omega).$$

For the proof, see Theorem 7.4 in [ACD:97]. More general tools are presented in [Ha:96]. We will improve the approximate differentiability, called in [Zi:89] the "L^1-differentiability", to the $L^{\frac{n}{n-1}}$-differentiability analogously as suggested in [AFP:00] for functions of bounded variation (Exercises 3.15-16). Let us mention that the proof of L^q-differentiability for $1 \leq q < \frac{n}{n-1}$ was recently done in [ABC:14].

We will need an appropriate version of the Poincaré-Korn inequality (see Remark II.1.1 of [Te:85] and Corollary 4.20 of [Bre:13]):

Theorem (7.5). *Let $\mathcal{R}$ be the set of all infinitesimal rigid motions in $\mathbb{R}^n$. For a bounded Lipschitz domain $\Omega \subset \mathbb{R}^n$, there exists a constant $C = C(\Omega)$ such that for every $u \in BD(\Omega; \mathbb{R}^n)$*

$$\min_{\rho \in \mathcal{R}} \|u - \rho\|_{L^{\frac{n}{n-1}}(\Omega; \mathbb{R}^n)} \leq C \|Eu\|_{M(\Omega; \mathbb{R}^{n \times n})}.$$

Therefore, for every projection R onto $\mathcal{R}$, we have (possibly for a larger constant)

$$\|u - R(u)\|_{L^{\frac{n}{n-1}}(\Omega; \mathbb{R}^n)} \leq C(R, \Omega) \|Eu\|_{M(\Omega; \mathbb{R}^{n \times n})}.$$

We will need this inequality explicitly for balls so let us according to Remark II.1.1 in [Te:85] define one projection. Fix $B_r(x_0) \subset \mathbb{R}^n$. Define

$$R_{x_0,r}(u)(x) := \frac{1}{|B_r|} \int_{B_r(x_0)} u(y) \, dy + \frac{1}{J_r} \left(\int_{B_r(x_0)} u(y) \times (y - x_0) \, dy \right)(x - x_0)$$

with J_r still to be defined. For the sake of simplicity, for $a, b \in \mathbb{R}^n$ we are denoting

$$a \times b := \tfrac{1}{2}(a \otimes b - b \otimes a).$$

Every $\rho \in \mathcal{R}$ can be written as a sum of a constant function, clearly invariant for $R_{x_0,r}$, and a linear combination of functions $x \mapsto (e_i \times e_j)(x - x_0)$. The average of such functions is 0. As for the second term, using the summation convention,

$$
\begin{aligned}
\left[(e_i \times e_j)y \times y\right]_{kl} &= \tfrac{1}{2}\left[(e_i \times e_j)y\right]_k y_l - \tfrac{1}{2}\left[(e_i \times e_j)y\right]_l y_k \\
&= \tfrac{1}{2}(e_i \times e_j)_{km} y_m y_l - \tfrac{1}{2}(e_i \times e_j)_{ln} y_n y_k \\
&= \tfrac{1}{4}(\delta_{ik}\delta_{jm} - \delta_{im}\delta_{jk}) y_m y_l - \tfrac{1}{4}(\delta_{il}\delta_{jn} - \delta_{in}\delta_{jl}) y_n y_k \\
&= \tfrac{1}{4}(\delta_{ik}y_j y_l - \delta_{jk}y_i y_l - \delta_{il}y_j y_k + \delta_{jl}y_i y_k).
\end{aligned}
$$

Therefore,

$$
\begin{aligned}
&\left[\int_{B_r(x_0)} [(e_i \times e_j)(y - x_0)] \times (y - x_0) \, dy \right]_{kl} \\
&= \int_{B_r(0)} \tfrac{1}{4}(\delta_{ik}y_j y_l - \delta_{jk}y_i y_l - \delta_{il}y_j y_k + \delta_{jl}y_i y_k) \, dy \\
&= \tfrac{1}{2}(\delta_{ik}\delta_{jl} - \delta_{jk}\delta_{il}) \int_{B_r(0)} y_1^2 \, dy \\
&= [e_i \times e_j]_{kl} \int_{B_r(0)} y_1^2 \, dy.
\end{aligned}
$$

Hence, by setting

$$J_r := \int_{B_r(0)} y_1^2 \, dy = \frac{\pi^{n/2}}{(n+2)\Gamma(\frac{n}{2}+1)} \, r^{n+2} =: J_1 r^{n+2}$$

to be the second moment of a ball about an axis through the origin, we achieve that also all maps $x \mapsto (e_i \times e_j)(x - x_0)$ are invariant under $R_{x_0,r}$, and consequently, $R_{x_0,r}$ is a projection.

By changing some signs in the computation above, we also see that this projection is in some sense orthogonal. Namely, all $x \mapsto [e_i \odot e_j](x - x_0)$ and therefore all maps induced by symmetric matrices lie in its kernel.

Now, let us denote by C_1 the constant in the Poincaré inequality for $B_1(0)$ with the corresponding projection, i.e., $C_1 = C(R_{0,1}, B_1(0))$. Take $u \in BD(B_r(x_0); \mathbb{R}^n)$, and let ω its rescaling to the unit ball, i.e.,

$$u(x_0 + r\xi) = \omega(\xi).$$

Then

$$\frac{1}{J_r} \int_{B_r(x_0)} u(x) \times (x - x_0) \, dx = \frac{1}{r^{n+2} J_1} \int_{B_1} u(x_0 + r\xi) \times r\xi \; r^n d\xi$$

$$= \frac{1}{r J_1} \int_{B_1} \omega(\xi) \times \xi \; d\xi$$

and so

$$R_{x_0,r}(u)(x_0 + r\xi) = R_{0,1}(\omega)(\xi).$$

By changing the variables in the Poincaré inequality, we see that the constant is translation and scaling invariant, i.e., $C(R_{x_0,r}, B_r(x_0)) = C(R_{0,1}, B_1(0))$.

Corollary (7.6). *Every $u \in BD(\Omega; \mathbb{R}^n)$ is a.e. $L^{\frac{n}{n-1}}$-differentiable and for a.e. $x_0 \in \Omega$ it holds*

$$\lim_{r \to 0} \frac{1}{r^n} \int_{B_r(x_0)} \left| \frac{u(x) - u(x_0) - \nabla u(x_0)(x - x_0)}{r} \right|^{\frac{n}{n-1}} dx = 0.$$

Proof. Let be $x_0 \in \Omega$ any point such that

- Theorem (7.4) holds,

- x_0 is Lebesgue point of $\mathfrak{E}u$,

- $\lim_{r \to 0} \frac{1}{r^n} |E^s u|(B_r(x_0)) = 0$.

By the Besicovitch derivation theorem (C.1), a.e. x_0 meets these conditions. Applying the Poincaré inequality to the function $\tilde{u}(x) := u(x) - u(x_0) - L_{x_0}(x - x_0)$ yields

$$\|\tilde{u} - R_{x_0,r}(\tilde{u})\|_{L^{\frac{n}{n-1}}} \le C_1 \|E\tilde{u}\|_M$$

or

$$\left(\frac{1}{r^n} \int_{B_r(x_0)} \frac{|\tilde{u}(x) - R_{x_0,r}(\tilde{u})(x)|^{\frac{n}{n-1}}}{r^{\frac{n}{n-1}}} dx \right)^{\frac{n-1}{n}} \le$$

$$\le \frac{C_1}{r^n} \left(\int_{B_r(x_0)} |\mathfrak{E}u(x) - \mathfrak{E}u(x_0)| \, dx + |E^s u|(B_r(x_0)) \right).$$

The right side converges for the chosen x_0 to 0 as $r \to 0$. Therefore, the claim will be proved when we show

$$\lim_{r \to 0} \frac{1}{r^n} \int_{B_r(x_0)} \frac{|R_{x_0,r}(\tilde{u})(x)|^{\frac{n}{n-1}}}{r^{\frac{n}{n-1}}} dx = 0.$$

Denote $\tilde{a}_r + \tilde{A}_r(x - x_0) := R_{x_0,r}(\tilde{u})(x)$. According to Theorem (7.4)

$$\lim_{r \to 0} \frac{\tilde{a}_r}{r} = \lim_{r \to 0} \frac{1}{|B_r(x_0)|} \int_{B_r(x_0)} \frac{\tilde{u}(x)}{r} \, dx = 0.$$

Moreover, from

$$\tilde{A}_r = \frac{1}{J_1 r^{n+2}} \int_{B_r(x_0)} \tilde{u}(x) \times (x - x_0) \, dx,$$

it follows

$$|\tilde{A}_r| \leq \frac{1}{J_1 r^{n+2}} \int_{B_r(x_0)} |\tilde{u}(x)||x - x_0| \, dx \leq \frac{1}{J_1 r^n} \int_{B_r(x_0)} \frac{|\tilde{u}(x)|}{r} \, dx \to 0.$$

Hence,

$$\lim_{r \to 0} \frac{1}{r^n} \int_{B_r(x_0)} \frac{|\tilde{A}_r(x - x_0)|^{\frac{n}{n-1}}}{r^{\frac{n}{n-1}}} \, dx = 0. \quad \blacksquare$$

7.1.4 Kernel of E_{dev}

Let us insert here a result that will not be necessary for our analysis but complements the theory. In the Poincaré-Korn inequality above, the kernel of E, which consists of all infinitesimal rigid motions, had to be excluded. In our considerations we split Eu into the deviatoric and the trace part. Actually even $E_{\mathrm{dev}}u := (Eu)_{\mathrm{dev}}$ contains enough information in order for the analogous Poincaré-Korn inequality to hold, see [Dai:06, FR:10]. Below we determine the corresponding kernel in an elementary way. Let us stress that the result has been known for a long time, and the proof may be found in Subsection 3.2.2 of [Re:94]. In the same manner, one gets an analogue of the Korn's inequality in the Sobolev spaces with $p > 1$, see, e.g., Theorem 3.3.2 in [Re:94].

Proposition (7.7). *Let $n \geq 3$. A smooth function $u : \Omega \to \mathbb{R}^n$ has trivial deviatoric part of the symmetrized gradient if and only if*

$$u(x) = (a \cdot x)x - \tfrac{1}{2}|x|^2 a + Wx + \beta x + b$$

for some vectors $a, b \in \mathbb{R}^n$, some $\beta \in \mathbb{R}$ and some skew-symmetric matrix $W \in \mathbb{R}^{n \times n}$.

For the sake of clarity, in the proof we denote the partial derivatives by $f_{,i} := \frac{\partial f}{\partial x_i}$.

Proof.

$\Rightarrow$:

Suppose $\mathfrak{E}_{\mathrm{dev}}u = 0$. Hence, $\mathfrak{E}u$ is in every point a scalar matrix, i.e. multiple of the identity matrix.

Step 1:

For $j \neq k$ we have

$$u_{j,k} + u_{k,j} = 0,$$

and therefore for every $i \neq k$ also

$$0 = u_{j,ki} + u_{k,ji} = u_{j,ik} + u_{k,ij} = u_{j,ik} - u_{i,kj} = (u_{j,i} - u_{i,j})_{,k}.$$

Hence $u_{j,i} - u_{i,j}$ only depends on x_i and x_j. For $i \neq j$ it holds $u_{j,i} + u_{i,j} = 0$, and consequently also $u_{i,j}$ depends just on x_i and x_j. Hence, every u_i has a form

$$u_i(x) = \sum_{j \neq i} f_{ij}(x_i, x_j)$$

for some smooth functions f_{ij}.

Step 2:

If we compare the diagonals, we notice that for any $i \neq j$

$$u_{i,i} = u_{j,j} = \tfrac{1}{n} \operatorname{div} u,$$

or with the denotation as above

$$\sum_{k \neq i} f_{ik,i} = \sum_{k \neq j} f_{jk,j}.$$

For $k \notin \{i, j\}$ therefore

$$f_{ik,ik} = f_{jk,jk}.$$

The left-hand side depends on x_i and x_k whereas the right one on x_j and x_k. Therefore, both are just some function of x_k. Hence, every f_{ik} has the form

$$f_{ik}(x_i, x_k) = x_i A_{ik}(x_k) + B_{ik}(x_i) + C_{ik}(x_k)$$

and for every $k \notin \{i, j\}$ also

$$A'_{ik} = A'_{jk}.$$

Therefore, we may suppose $A_{ik} = A_{jk} =: A_k$ and take the possible constant difference into account in the functions B_{ik}. Hence, by defining $B_i := \sum_{k \neq i} B_{ik}$,

$$u_i(x) = \sum_{k \neq i} f_{ik}(x_i, x_k) = \sum_{k \neq i} A_k(x_k)x_i + B_i(x_i) + \sum_{k \neq i} C_{ik}(x_k).$$

Step 3:

Now reconsider $u_{i,j} = -u_{j,i}$ for $i \neq j$. Then

$$x_i A'_j(x_j) + C'_{ij}(x_j) = -(x_j A'_i(x_i) + C'_{ji}(x_i)).$$

Differentiating the equation with respect to x_i and x_j yields

$$A''_j(x_j) = -A''_i(x_i).$$

By taking distinct i, j and k we obtain

$$A''_k(x_k) = -A''_i(x_i) = A''_j(x_j) = -A''_k(x_k).$$

Therefore, $A''_k = 0$, and A_k is a linear function for every k. We may suppose $A_k(x_k) = a_k x_k$ for some $a_k \in \mathbb{R}$ since in the expression for u_i the constant term would yield a linear function in x_i and we may regard it as part of the so far undetermined function B_i. The equation $u_{i,j} = -u_{j,i}$ now reads

$$a_j x_i + C'_{ij}(x_j) = -(a_i x_j + C'_{ji}(x_i)).$$

By taking derivative with respect to x_j, it follows

$$C''_{ij}(x_j) = -a_i.$$

Therefore, again by moving the possible constant to B_i, we get

$$C_{ij}(x_j) = -\tfrac{1}{2}a_i x_j^2 + w_{ij} x_j$$

for some $w_{ij} \in \mathbb{R}$. From the equation above it also follows that $w_{ji} = -w_{ij}$.

Step 4:

We have got the representation

$$u_i(x) = \sum_{k \neq i} a_k x_k x_i + B_i(x_i) + \sum_{k \neq i} (-\tfrac{1}{2} a_i x_k^2 + w_{ik} x_k).$$

We may complete the sums by adding the missing i-th term since they again depend only on x_i. The corresponding subtraction may be regarded as part of B_i. By introducing a vector a with components a_i and a skew-symmetric matrix W with entries w_{ij}, we obtain

$$u_i(x) = (a \cdot x) x_i + B_i(x_i) - \tfrac{1}{2} a_i |x|^2 + (Wx)_i.$$

Step 5:

Now we again consider the condition on the diagonal. Since

$$u_{i,i}(x) = a_i x_i + a \cdot x + B_i'(x_i) - a_i x_i = a \cdot x + B_i'(x_i),$$

and $u_{i,i} = u_{j,j}$, it follows

$$B_i'(x_i) = B_j'(x_j) =: \beta \in \mathbb{R}.$$

Therefore, there is some $b \in \mathbb{R}^n$ such that

$$B_i(x_i) = \beta x_i + b_i.$$

Recapitulation:

We have proved that u must have the following form

$$u_i(x) = (a \cdot x) x_i - \tfrac{1}{2} a_i |x|^2 + \beta x_i + b_i + (Wx)_i$$

or in vector form

(7.8) $$u(x) = (a \cdot x) x - \tfrac{1}{2} |x|^2 a + Wx + \beta x + b.$$

$\Leftarrow$:

Now suppose that u has the form from (7.8). From

$$\nabla u(x) = x \otimes a + (a \cdot x) I - a \otimes x + W + \beta I$$

it follows

$$\mathfrak{E} u(x) = (a \cdot x + \beta) I. \quad \blacksquare$$

Remark (7.9). In some steps of the proof, three different indices were employed. Therefore, the assumption $n \geq 3$ was really necessary. The kernel is indeed different in the remaining two cases. Obviously, the case $n = 1$ is trivial. If $n = 2$, the condition reads

$$
\begin{aligned}
0 &= \mathfrak{E}_{\mathrm{dev}} u \\
&= \begin{bmatrix} u_{1,1} & \tfrac{1}{2}(u_{1,2} + u_{2,1}) \\ \tfrac{1}{2}(u_{1,2} + u_{2,1}) & u_{2,2} \end{bmatrix} - \frac{1}{2} \begin{bmatrix} u_{1,1} + u_{2,2} & 0 \\ 0 & u_{1,1} + u_{2,2} \end{bmatrix} \\
&= \frac{1}{2} \begin{bmatrix} u_{1,1} - u_{2,2} & u_{1,2} + u_{2,1} \\ u_{1,2} + u_{2,1} & u_{2,2} - u_{1,1} \end{bmatrix}.
\end{aligned}
$$

Therefore

$$0 = \mathfrak{E}_{\mathrm{dev}} u \iff u_{1,1} = u_{2,2} \text{ and } u_{1,2} = -u_{2,1}.$$

The latter is exactly the Cauchy-Riemann system. Hence, u lies in the kernel if and only if it is holomorphic, i.e., the function $x_1 + i x_2 \mapsto u_1(x_1, x_2) + i u_2(x_1, x_2)$ is holomorphic on $\{x_1 + i x_2 : (x_1, x_2) \in \Omega\} \subset \mathbb{C}$.

7.2   Space U

7.2.1   Definition and basic properties

For an integral functional whose density has Hencky plasticity growth, its natural domain is the space

$$LU(\Omega; \mathbb{R}^n) := \{u \in LD(\Omega; \mathbb{R}^n) : \operatorname{div} u \in L^2(\Omega)\}.$$

Endowed with

$$\|u\|_{U(\Omega; \mathbb{R}^n)} := \|u\|_{LD(\Omega; \mathbb{R}^n)} + \|\operatorname{div} u\|_{L^2(\Omega)},$$

it is clearly a Banach space. If Ω is a bounded Lipschitz domain, $C^\infty(\overline{\Omega}; \mathbb{R}^n)$ is a dense subset (combine Proposition I.1.3 with the proof of Theorem II.3.4 in [Te:85]). We will analogously as in Sobolev spaces denote

$$LU_0(\Omega; \mathbb{R}^n) := \overline{C_c^\infty(\Omega; \mathbb{R}^n)}.$$

Let $f : \Omega \times \mathbb{R}^{n \times n}_{\text{sym}} \to \mathbb{R}$ be a Carathéodory function with Hencky plasticity growth. For every $X \in \mathbb{R}^{n \times n}_{\text{sym}}$, $\varphi \in LU_0(\Omega; \mathbb{R}^n)$ and $\varepsilon > 0$, there exists $\varphi_\varepsilon \in C_c^\infty(\Omega; \mathbb{R}^n)$ such that

$$\int_\Omega f(x, X + \mathfrak{E}\varphi(x)) \, dx \geq \int_\Omega f(x, X + \mathfrak{E}\varphi_\varepsilon(x)) - \varepsilon.$$

The proof follows the usual scheme of employing Fatou's lemma, passing to a.e. pointwisely convergent sequence and using continuity of f in the second variable.

Due to the lack of weak compactness of bounded sequences in the space LU, we introduce the corresponding space

$$U(\Omega; \mathbb{R}^n) := \{u \in BD(\Omega; \mathbb{R}^n) : \operatorname{div} u \in L^2(\Omega)\}$$

(with the obvious norm). Clearly, since the trace part of Eu is regular, we have $E^s_{\text{dev}} u = E^s u$. The definitions and claims for BD may be adapted in the following manner.

Definition (7.10). Let us have $\{u_j\}_{j \in \mathbb{N}} \subset U(\Omega; \mathbb{R}^n)$ and $u \in U(\Omega; \mathbb{R}^n)$.

(a) We say that $\{u_j\}_{j \in \mathbb{N}}$ *weakly converges* to u in $U(\Omega; \mathbb{R}^n)$, $u_j \rightharpoonup u$, if

- $u_j \to u$ in $L^1(\Omega; \mathbb{R}^n)$,
- $Eu_j \overset{*}{\rightharpoonup} Eu$ in $M(\Omega; \mathbb{R}^{n \times n})$,
- $\operatorname{div} u_j \rightharpoonup \operatorname{div} u$ in $L^2(\Omega)$.

(b) The *strict* (or *intermediate*) *convergence*, $u_j \overset{|\cdot|}{\to} u$, means that

- $u_j \to u$ in $L^1(\Omega; \mathbb{R}^n)$,
- $|Eu_j|(\Omega) \to |Eu|(\Omega)$,
- $\operatorname{div} u_j \to \operatorname{div} u$ in $L^2(\Omega)$.

(The underlying metric is clearly

$$d(u, v) := \|u - v\|_{L^1} + \big||Eu|(\Omega) - |Ev|(\Omega)\big| + \|\operatorname{div} u - \operatorname{div} v\|_{L^2}.)$$

(c) Let $c : \mathbb{R}_{\mathrm{sym}}^{n \times n} \to \mathbb{R}$ be a non-negative convex function with linear upper bound. Then $\{u_j\}_{j \in \mathbb{N}}$ *converges c-strictly* in $U(\Omega; \mathbb{R}^n)$ to u, symbolically $u_j \overset{c}{\rightharpoonup} u$, if

- $u_j \overset{|\cdot|}{\rightharpoonup} u$ in $U(\Omega; \mathbb{R}^n)$,
- $\int_\Omega c(E_{\mathrm{dev}} u_j) \to \int_\Omega c(E_{\mathrm{dev}} u)$,
- $\int_\Omega c(E u_j) \to \int_\Omega c(E u)$.

First let us mention that a bounded sequence from $U(\Omega; \mathbb{R}^n)$ contains a weakly convergent subsequence. This follows immediately from the corresponding results in $BD(\Omega; \mathbb{R}^n)$ and $L^2(\Omega)$.

For functions in $U(\Omega; \mathbb{R}^n)$ outside $LU(\Omega; \mathbb{R}^n)$, an approximation by smooth functions is not possible in the norm topology. However, we may at least get an approximation in the c-strict topology. We give this result in the form of Theorem 14.1.4 from [ABM:06]:

Theorem (7.11). *Let $\Omega \subset \mathbb{R}^n$ be a bounded Lipschitz domain and $c : \mathbb{R}_{\mathrm{sym}}^{n \times n} \to \mathbb{R}$ a non-negative convex function such that*

- *there exist $\alpha, \beta > 0$ such that for all $X \in \mathbb{R}_{\mathrm{sym}}^{n \times n}$ it holds*

$$\alpha(|X| - 1) \le c(X) \le \beta(|X| + 1),$$

- *the domain of its conjugate c^* is closed.*

Then for every $u \in U(\Omega; \mathbb{R}^n)$ there exists $\{u_j\}_{j \in \mathbb{N}} \subset C^\infty(\Omega; \mathbb{R}^n) \cap LU(\Omega; \mathbb{R}^n)$ such that

$$\gamma(u_j) = \gamma(u) \quad and \quad u_j \overset{c}{\rightharpoonup} u \text{ in } U(\Omega; \mathbb{R}^n).$$

Remark (7.12). In Section 8.3 we will use this approximation with the convex function $c = \langle \cdot \rangle$ defined as

$$\langle X \rangle := \sqrt{1 + |X|^2}.$$

It obviously fulfils the growth conditions and $\langle \cdot \rangle^*$ has closed domain since

$$\langle Y \rangle^* = \begin{cases} -\sqrt{1 - |Y|^2}, & |Y| \le 1, \\ \infty, & |Y| > 1. \end{cases}$$

7.2.2 Helmholtz decomposition

In L^p-spaces for $p > 1$, there exists a form of the Helmholtz decomposition (e.g., Section 2.3 in [Kr:94] or Example 3.14 in [Gr:90]). We show that a similar result holds also in the space U. We will need the following standard existence and regularity result for Poisson's equation.

Theorem (7.13). *Let $\Omega \subset \mathbb{R}^n$ be a cube or a bounded open set with $C^{1,1}$-boundary. For any $f \in L^2(\Omega)$ the Dirichlet problem for Poisson's equation*

$$\triangle \phi = f, \quad \phi \in W_0^{1,2}(\Omega),$$

has a unique weak solution $\phi \in W^{2,2}(\Omega)$. Moreover, there exists $C > 0$ such that

$$\|\phi\|_{W^{2,2}} \le C \|f\|_{L^2}.$$

For the proof we refer to Theorem 9.15 in [GT:01] for $C^{1,1}$-domains and to Section 9.1 in [WY:06] for cubes.

Proposition (7.14). *Let $\Omega \subset \mathbb{R}^n$ be a bounded open set with $C^{1,1}$-boundary or a cube. Then for every $u \in U(\Omega; \mathbb{R}^n)$ there exist unique*

$$v \in U(\Omega; \mathbb{R}^n) \text{ with } \operatorname{div} v = 0 \quad and \quad \phi \in W_0^{1,2}(\Omega) \cap W^{2,2}(\Omega)$$

such that $u = v + \nabla \phi$. Therefore, we have a decomposition

$$U(\Omega; \mathbb{R}^n) = (\ker \operatorname{div}) \oplus (\operatorname{im} \nabla)$$

into two closed subspaces where here

$$\nabla : W_0^{1,2}(\Omega) \cap W^{2,2}(\Omega) \to W^{1,2}(\Omega; \mathbb{R}^n).$$

Proof. Let us define a map

$$P : U(\Omega; \mathbb{R}^n) \to U(\Omega; \mathbb{R}^n), \quad P(u) := \nabla \phi$$

with $\phi \in W^{2,2}(\Omega)$ being the unique weak solution of

$$\triangle \phi = \operatorname{div} u, \quad \phi \in W_0^{1,2}(\Omega).$$

P is linear and idempotent. According to Theorem (7.13), also

$$\|\nabla \phi\|_U = \|\nabla \phi\|_{L^1} + \|\nabla \nabla \phi\|_M + \|\triangle \phi\|_{L^2} \leq C_1 \|\phi\|_{W^{2,2}} \leq C_2 \|\operatorname{div} u\|_{L^2} \leq C_2 \|u\|_U.$$

Therefore, P is a projection. By the definition, $\operatorname{im} P \subset \nabla\big(W_0^{1,2}(\Omega) \cap W^{2,2}(\Omega)\big)$. Since for every $\phi \in W_0^{1,2}(\Omega) \cap W^{2,2}(\Omega)$ it holds $P(\nabla \phi) = \nabla \phi$, actually

$$\operatorname{im} P = \nabla\big(W_0^{1,2}(\Omega) \cap W^{2,2}(\Omega)\big).$$

Moreover, from

$$\operatorname{div} u = 0 \iff \nabla \phi = 0 \iff \phi = 0,$$

it follows that $\ker P = \ker \operatorname{div}$.   ∎

According to Lemma (C.4), for bounded sequences in $W^{1,p}$, $p > 1$, there exists a modified sequence with p-equiintegrable gradients. For a function from $U(\Omega; \mathbb{R}^n)$, only a part of the symmetrized gradient has a higher integrability. Still, we may get a similar result where we achieve 2-equiintegrability in that part.

Lemma (7.15). *Let $\Omega \subset \mathbb{R}^n$ be an open bounded set with $C^{1,1}$-boundary and let $\{u_j\}_{j \in \mathbb{N}}$ be a bounded sequence in $U(\Omega; \mathbb{R}^n)$. There exist a subsequence $\{u_{j_k}\}_{k \in \mathbb{N}}$ and a sequence $\{\tilde{u}_k\}_{k \in \mathbb{N}} \subset U(\Omega; \mathbb{R}^n)$ such that*

- $\{(\operatorname{div} \tilde{u}_k)^2\}_{k \in \mathbb{N}}$ *is equiintegrable,*

- $\{u_{j_k} - \tilde{u}_k\}_{k \in \mathbb{N}} \subset W^{1,2}(\Omega; \mathbb{R}^n)$ *and therefore $E^s u_{j_k} = E^s \tilde{u}_k$,*

- $\lim_{k \to \infty} \big|\{\nabla(\tilde{u}_k - u_{j_k}) \neq 0\} \cup \{\tilde{u}_k \neq u_{j_k}\}\big| = 0$.

Moreover, if $\{u_j\}_{j \in \mathbb{N}}$ converges weakly, strictly or c-strictly to u in $U(\Omega; \mathbb{R}^n)$, then the $\tilde{u}_k$ can be chosen in such a way that $\gamma(\tilde{u}_k) = \gamma(u)$ and $\{\tilde{u}_k\}_{k \in \mathbb{N}}$ converges to u in $U(\Omega; \mathbb{R}^n)$ in the same manner.

Proof. Let us decompose

$$u_j = v_j + \nabla \phi_j$$

according to Proposition (7.14). Since $\triangle \phi_j = \operatorname{div} u_j$, $\{\phi_j\}_{j \in \mathbb{N}}$ is a bounded sequence in $W^{2,2}(\Omega)$ by Theorem (7.13). Denote $w_j := \nabla \phi_j$. A suitable subsequence of $\{w_j\}_{j \in \mathbb{N}}$ converges weakly in $W^{1,2}(\Omega; \mathbb{R}^n)$ to some w. According to the Lemma (C.4), there exist a further subsequence $\{w_{j_k}\}_{k \in \mathbb{N}}$ and a sequence $\{\tilde{w}_k\}_{k \in \mathbb{N}} \subset w + W_0^{1,2}(\Omega; \mathbb{R}^n)$ such that

- $\tilde{w}_k \rightharpoonup w$ in $W^{1,2}(\Omega; \mathbb{R}^n)$,

- $\{|\nabla \tilde{w}_k|^2\}_{k \in \mathbb{N}}$ is equiintegrable,

- $\lim_{k \to \infty} |\{w_{j_k} \neq \tilde{w}_k\} \cup \{\nabla w_{j_k} \neq \nabla \tilde{w}_k\}| = 0$.

Define

$$\tilde{u}_k := v_{j_k} + \tilde{w}_k.$$

It is a bounded sequence in $U(\Omega; \mathbb{R}^n)$ that has the desired properties.

As for supplement: If $u_j \rightharpoonup u$, then also the related projections weakly converge, and we continue as above. For the $(c\text{-})$strict convergence we assess, employing Lipschitz continuity of c,

$$
\begin{aligned}
\left| \int_\Omega c(Eu_{j_k}) - \int_\Omega c(E\tilde{u}_k) \right| &= \left| \int_\Omega \left(c(\mathfrak{E}u_{j_k}(x)) - c(\mathfrak{E}\tilde{u}_k(x)) \right) dx \right| \\
&\leq \int_\Omega L \left| \mathfrak{E}u_{j_k}(x) - \mathfrak{E}\tilde{u}_k(x) \right| dx \\
&= \int_{\{\mathfrak{E}w_{j_k} \neq \mathfrak{E}\tilde{w}_k\}} L \left| \mathfrak{E}w_{j_k}(x) - \mathfrak{E}\tilde{w}_k(x) \right| dx.
\end{aligned}
$$

If $k \to \infty$, the last expression converges to 0 since $\{\mathfrak{E}w_{j_k} - \mathfrak{E}\tilde{w}_k\}_{k \in \mathbb{N}}$ is bounded in L^2 and is thus equiintegrable. Similarly for E_{dev}. ∎

Remark (7.16). Young measures offer a possibility to describe weak convergence more precisely. Namely, a highly oscillatory sequence may converge to a constant, which clearly does not contain any information about the members. For integral functionals that we explore, gradient (or better symmetrized-gradient) Young measures should be considered. While sufficient for $p > 1$, the theory must be generalized for $p = 1$ since beside oscillations also concentrations must be incorporated. The concept dates back to the article [DM:87] from DiPerna and Majda where they cope with a specific problem. The frame-work for the general theory was set in [AB:97]. For the full-gradient case, the corresponding generalized gradient Young measures generated by sequences in BV were identified in [KR:10-2]. Recently, also the symmetrized-gradient case was resolved, see [DR:17]. The right analogue concept for the Hencky plasticity would contain measures generated by sequences in U. By the decomposition lemma (7.15), the generating sequence may be taken to have 2-equiintegrable divergences.

7.3   Bogovskii's operator

In the next chapter we will have to impose the right boundary values on the functions from $LU(\Omega; \mathbb{R}^n)$. Usually, it is done by employing a suitable cut-off function. Here we will have to pay more attention since we may land outside $U(\Omega; \mathbb{R}^n)$. Namely, for $\varphi \in C_c^\infty(\Omega)$ and $v \in LU(\Omega; \mathbb{R}^n)$, the function

$$\operatorname{div}(\varphi v) = \nabla\varphi \cdot v + \varphi \operatorname{div} v$$

lies a priori merely in $L^{\frac{n}{n-1}}(\Omega; \mathbb{R}^n)$ due to the first summand. We will fix this by adding a corrector term that will provide the necessary integrability. Thus, we will have an additional term also in the remainder of the symmetrized gradient that must, however, lie only in L^1. Our tool is given by the next theorem.

Theorem (7.17). *Let $\Omega \subset \mathbb{R}^n$ be a bounded Lipschitz domain and $1 < q < \infty$. There exists a linear operator $\mathcal{B} = \mathcal{B}_{\Omega,q} : L^q(\Omega) \to W_0^{1,q}(\Omega; \mathbb{R}^n)$ with the following properties:*

- *For every $f \in L^q(\Omega)$ with $\int_\Omega f(x)\, dx = 0$, it holds*

$$\operatorname{div} \mathcal{B}f = f.$$

- *For every $f \in L^q(\Omega)$*

$$\|\nabla(\mathcal{B}f)\|_{L^q(\Omega; \mathbb{R}^{n \times n})} \leq C\|f\|_{L^q(\Omega)}.$$

The constant C depends only on Ω and q, and is translation- and scaling-invariant.

- *If $f \in C_c^\infty(\Omega)$, then $\mathcal{B}f \in C_c^\infty(\Omega; \mathbb{R}^n)$.*

In this theorem we gathered the relevant results from Section 2 in [BS:90]. See also the references therein, as the original proofs go back to Bogovskii. Therefore, $\mathcal{B}$ is sometimes referred to as Bogovskii's operator.

Now we return to our discussion. Let us fix $1 < q \leq \frac{n}{n-1}$ (with $n \geq 2$).

Let $B \subset \Omega$ be an open set that contains $\operatorname{supp} \nabla\varphi$. By Theorem (7.2), $v \in L^q(\Omega; \mathbb{R}^n)$, and we may define $z := \mathcal{B}_{B,q}f \in W_0^{1,q}(B; \mathbb{R}^n)$ where

$$f := -\nabla\varphi \cdot v + \frac{1}{|B|}\int_B \nabla\varphi(x) \cdot v(x)\, dx \in L^q(B).$$

Since $\operatorname{div} z = f$ in B,

$$\varphi v + z \in LU(\Omega; \mathbb{R}^n)$$

where we as usual extended z by the zero function. Let us show that this function lies even in $LU_0(\Omega; \mathbb{R}^n)$. We will use a similar technique as in [AG:82].

There exists $\{v_k\}_{k \in \mathbb{N}} \subset C^\infty(\overline{\Omega}; \mathbb{R}^n)$ such that

$$\|v_k - v\|_{LU(\Omega; \mathbb{R}^n)} \to 0, \quad \text{and therefore also} \quad \|v_k - v\|_{L^q(\Omega; \mathbb{R}^n)} \to 0.$$

We choose a sequence $\{\vartheta_k\}_{k \in \mathbb{N}} \in C_c^\infty(B)$ such that

$$\vartheta_k \to \frac{1}{|B|}\int_B \nabla\varphi(x) \cdot v(x)\, dx \quad \text{in } L^2(B).$$

If $\int_B \nabla\varphi(x) \cdot v(x) \, dx = 0$, we take $\vartheta_k := \frac{1}{k}\vartheta$ for some non-negative nontrivial $\vartheta \in C_c^\infty(B)$. Thus (at least for k large enough) we may define

$$f_k := -\nabla\varphi \cdot v_k + \vartheta_k \frac{\int_B \nabla\varphi(x) \cdot v_k(x) \, dx}{\int_B \vartheta_k(x) \, dx} \in C_c^\infty(B).$$

Then $z_k := \mathcal{B}_B f_k \in C_c^\infty(B; \mathbb{R}^n)$, and $\operatorname{div} z_k = f_k$. Since

$$\lim_{k\to\infty} \int_B \nabla\varphi(x) \cdot v_k(x) \, dx = \int_B \nabla\varphi(x) \cdot v(x) \, dx = \lim_{k\to\infty} \int_B \vartheta_k(x) \, dx,$$

also

$$\vartheta_k \frac{\int_B \nabla\varphi(x) \cdot v_k(x) \, dx}{\int_B \vartheta_k(x) \, dx} \to \frac{1}{|B|} \int_B \nabla\varphi(x) \cdot v(x) \, dx \quad \text{in } L^2(B).$$

(This holds also for the trivial case.) Hence,

$$\|\nabla z_k - \nabla z\|_{L^q(\Omega;\mathbb{R}^{n\times n})} \le C\|f_k - f\|_{L^q(B)} \to 0.$$

By the Poincaré inequality also $\|z_k - z\|_{L^q(\Omega;\mathbb{R}^n)} \to 0$. Thus $\varphi v_k + z_k \in C_c^\infty(\Omega; \mathbb{R}^n)$, and

$$\|(\varphi v_k + z_k) - (\varphi v + z)\|_{LD(\Omega;\mathbb{R}^n)} \to 0.$$

Moreover, since on Ω

$$\operatorname{div}(\varphi v_k + z_k) = \varphi \operatorname{div} v_k + \vartheta_k \frac{\int_B \nabla\varphi(x) \cdot v_k(x) \, dx}{\int_B \vartheta_k(x) \, dx},$$

$$\operatorname{div}(\varphi v + z) = \varphi \operatorname{div} v + 1_B \frac{1}{|B|} \int_B \nabla\varphi(x) \cdot v(x) \, dx,$$

also

$$\|\operatorname{div}(\varphi v_k + z_k) - \operatorname{div}(\varphi v + z)\|_{L^2(\Omega)} \to 0.$$

Chapter 8

Commutability of homogenization and vanishing hardening

Having made the necessary preparations, we are now in the position to investigate our problem. First, we will repeat our setting clearly. We will extend the notion of homogenization to the functions with Hencky plasticity growth. Some properties still remain valid in spite of the non-standard growth.

Then, we will focus on the homogenization of the energy functional for the zero-hardening case, which is the core of Part II. We will start by constructing a recovery sequence, following the ideas in [KR:10-1]. The sequence will converge in the $\langle \cdot \rangle$-strict topology so that we will be allowed to apply the Reshetnyak continuity theorem. We will also need Theorem (7.3), the analogue of Alberti's rank-one theorem for functions of bounded deformation. Besides, we will use the known results for the densities with linear growth. We will have to handle the recession function with special attention. For that purpose we will show a rather technical assessment on the Lipschitz constant.

Regarding the lim inf-inequality, the approach from [ABM:06] with the slicing method of De Giorgi can be adapted for the regular points. As in [DQ:90], we will have to take care of the divergence. The tool will be Theorem (7.17), and therefore, we will have to move the analysis form L^1 to L^q. By Corollary (7.6) it is still possible to employ approximate differentiability. To control also the singular points, we will suppose the density to behave for large arguments as a convex function. This asymptotic convexity will enable the use of the results from [DQ:90]. Let us stress that for the finite arguments, however, we do not assume any convexity.

In Section 8.5 we will gather the results into two main theorems. Theorem (8.22) will cover the homogenization at zero hardening, or purely mathematically speaking, the homogenizability of functions that are $\mathbb{I}^n$-periodic in the first variable and have Hencky plasticity growth. With Theorem (8.23) we will give a positive answer to the question of commutability of homogenization and vanishing hardening.

Lastly, we will re-examine a special case: the relaxation problem. In [KK:16] new results regarding automatic convexity of 1-homogeneous functions with enough convex directions were shown. Instead of the asymptotic convexity, we will impose a reasonable growth condition and show an analogous relaxation result. We will again employ the analogue to Alberti's rank-one theorem.

8.1   Setting

First, let us concisely repeat the setting and explain the known facts. Throughout the chapter we will consider a Carathéodory function $f : \mathbb{R}^n \times \mathbb{R}^{n \times n}_{\text{sym}} \to \mathbb{R}$ that is

- $\mathbb{I}^n$-periodic in the first variable,

- has growth properties typical for the densities in the Hencky plasticity, i.e., there exist $\alpha, \beta > 0$ such that for all $x \in \Omega$ and $X \in \mathbb{R}^{n \times n}_{\text{sym}}$

$$\alpha(|X_{\text{dev}}| + (\text{tr}\, X)^2) \leq f(x, X) \leq \beta(|X_{\text{dev}}| + (\text{tr}\, X)^2 + 1).$$

For any $\delta \geq 0$ we define

$$f^{(\delta)} : \mathbb{R}^n \times \mathbb{R}^{n \times n}_{\text{sym}} \to \mathbb{R}, \quad f^{(\delta)}(x, X) := f(x, X) + \delta |X_{\text{dev}}|^2.$$

We will investigate the integral functionals $\mathcal{F}_\varepsilon$ and $\mathcal{F}_\varepsilon^{(\delta)}$, $\delta \geq 0$, on $L^1(\Omega; \mathbb{R}^n)$ defined as

$$\mathcal{F}_\varepsilon(u) := \begin{cases} \int_\Omega f(\frac{x}{\varepsilon}, \mathfrak{E}u(x))\, dx, & u \in LU(\Omega; \mathbb{R}^n), \\ \infty, & \text{else,} \end{cases}$$

and

$$\mathcal{F}_\varepsilon^{(\delta)}(u) := \begin{cases} \int_\Omega f^{(\delta)}(\frac{x}{\varepsilon}, \mathfrak{E}u(x))\, dx, & u \in W^{1,2}(\Omega; \mathbb{R}^n), \\ \infty, & \text{else.} \end{cases}$$

Obviously, for any $u \in L^1(\Omega; \mathbb{R}^n)$ we have

$$\mathcal{F}_\varepsilon^{(\delta)}(u) \searrow \mathcal{F}_\varepsilon^{(0)}(u) \quad \text{as} \quad \delta \to 0.$$

Then also

$$\Gamma(L^1)\text{-}\lim_{\delta \to 0} \mathcal{F}_\varepsilon^{(\delta)} = \text{lsc}\, \mathcal{F}_\varepsilon^{(0)} = \text{lsc}\, \mathcal{F}_\varepsilon.$$

The last equality follows from the fact that $W^{1,2}(\Omega; \mathbb{R}^n)$, even $C^\infty(\overline{\Omega}; \mathbb{R}^n)$, is dense in $(LU(\Omega; \mathbb{R}^n), \|.\|_U)$, and can be proved by the same strategy as the one at the beginning of Subsection 7.2.1.

For $\delta > 0$ we may apply Proposition (2.29) and Corollary (2.33). Hence, for every $u \in L^2(\Omega; \mathbb{R}^n)$

$$\Gamma(L^2)\text{-}\lim_{\varepsilon \to 0} \mathcal{F}_\varepsilon^{(\delta)}(u) = \mathcal{F}_{\text{hom}}^{(\delta)}(u)$$

where

$$\mathcal{F}_{\text{hom}}^{(\delta)}(u) := \begin{cases} \int_\Omega f_{\text{hom}}^{(\delta)}(\mathfrak{E}u(x))\, dx, & u \in W^{1,2}(\Omega; \mathbb{R}^n), \\ \infty, & \text{else,} \end{cases}$$

has density

$$f_{\text{hom}}^{(\delta)}(X) = \inf_{k \in \mathbb{N}} \inf_{\varphi \in W_0^{1,2}(k\mathbb{I}^n; \mathbb{R}^n)} \frac{1}{k^n} \int_{k\mathbb{I}^n} f^{(\delta)}(x, X + \mathfrak{E}\varphi(x))\, dx.$$

This also holds with respect to L^1-norm (on the whole $L^1(\Omega; \mathbb{R}^n)$) due to the lower bound and Korn's inequality. Schematically, so far we have

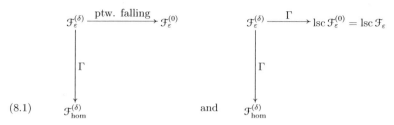

(8.1)

8.2 Homogenized density

We expect the density of the homogenized functional to have an analogous form also for f (of course, with an additional singular term). Therefore, we define

$$f_{\text{hom}}(X) := \inf_{k \in \mathbb{N}} \inf_{\varphi \in C_c^\infty(k\mathbb{I}^n; \mathbb{R}^n)} \frac{1}{k^n} \int_{k\mathbb{I}^n} f(x, X + \mathfrak{E}\varphi(x)) \, dx.$$

By the reasoning in Subsection 7.2.1,

$$f_{\text{hom}}(X) = \inf_{k \in \mathbb{N}} \inf_{\varphi \in LU_0(k\mathbb{I}^n; \mathbb{R}^n)} \frac{1}{k^n} \int_{k\mathbb{I}^n} f(x, X + \mathfrak{E}\varphi(x)) \, dx.$$

Let us explore the properties of the new function. We rely on the theory on subadditive processes that we employed already in Chapter 4. Here, however, we deal even with the invariant (i.e. non-stochastic) case.

For $X \in \mathbb{R}^{n \times n}_{\text{sym}}$ and $A \in \mathcal{B}_b(\mathbb{R}^n)$ let us define

$$m_X(A) := \inf_{\varphi \in C_c^\infty(\text{Int } A; \mathbb{R}^n)} \int_{\text{Int } A} f(x, X + \mathfrak{E}\varphi(x)) \, dx.$$

Then the mapping

$$m_X : \mathcal{B}_b(\mathbb{R}^n) \to \mathbb{R}$$

is subadditive and covariant (as in Proposition (4.6)). Moreover, for the spacial constant

$$\gamma(m_X) = \inf_{I \in \mathbb{Z}^n, \, |I| > 0} \frac{1}{|I|} \inf_{\varphi \in C_c^\infty(\text{Int } I; \mathbb{R}^n)} \int_{\text{Int } I} f(x, X + \mathfrak{E}\varphi(x)) \, dx,$$

it holds

$$0 \le \gamma(m_X) \le \beta(|X_{\text{dev}}| + (\text{tr } X)^2 + 1).$$

According to Theorem 2.1 from [LM:02], for every open bounded convex set A and $\varepsilon_k \searrow 0$

$$\gamma(m_X) = \lim_{k \to \infty} \inf \left\{ \frac{1}{|\varepsilon_k^{-1} A|} \int_{\varepsilon_k^{-1} A} f(x, X + \mathfrak{E}\varphi(x)) \, dx : \varphi \in C_c^\infty(\varepsilon_k^{-1} A; \mathbb{R}^n) \right\}$$

$$= f_{\text{hom}}(X).$$

For the functions with standard p-growth, the homogenized version is quasiconvex and is the same for the quasiconvex envelope. While in the proofs of these facts we may refer to more general results on Γ-convergence and relaxation, here we rely on regularization by hardening. Clearly, $f_{\mathrm{hom}}^{(\delta)}(X)$ is descending as $\delta \searrow 0$. Moreover,

$$
\begin{aligned}
\inf_{\delta>0} f_{\mathrm{hom}}^{(\delta)}(X) &= \inf_{\delta>0} \inf_{k\in\mathbb{N}} \inf_{\varphi\in C_c^\infty(k\mathbb{I}^n;\mathbb{R}^n)} \frac{1}{k^n} \int_{k\mathbb{I}^n} f^{(\delta)}(x, X + \mathfrak{E}\varphi(x))\, dx \\
&= \inf_{k\in\mathbb{N}} \inf_{\varphi\in C_c^\infty(k\mathbb{I}^n;\mathbb{R}^n)} \inf_{\delta>0} \frac{1}{k^n} \int_{k\mathbb{I}^n} f^{(\delta)}(x, X + \mathfrak{E}\varphi(x))\, dx \\
&= \inf_{k\in\mathbb{N}} \inf_{\varphi\in C_c^\infty(k\mathbb{I}^n;\mathbb{R}^n)} \frac{1}{k^n} \int_{k\mathbb{I}^n} f(x, X + \mathfrak{E}\varphi(x))\, dx \\
&= f_{\mathrm{hom}}(X),
\end{aligned}
$$

where we employed the dominated convergence theorem (for arbitrary sequence $\delta_j \searrow 0$). Hence, again applying the same theorem, the function f_{hom} is symmetric-quasiconvex.

Let us gather the results (and additional properties).

Proposition (8.2). *Let $f : \mathbb{R}^n \times \mathbb{R}_{\mathrm{sym}}^{n\times n} \to \mathbb{R}$ be a Carathéodory function that is $\mathbb{I}^n$-periodic in the first variable and fulfils for some $0 < \alpha < \beta$*

$$
\alpha(|X_{\mathrm{dev}}| + (\mathrm{tr}\, X)^2) \le f(x, X) \le \beta(|X_{\mathrm{dev}}| + (\mathrm{tr}\, X)^2 + 1)
$$

for a.e. $x \in \mathbb{R}^n$ and all $X \in \mathbb{R}_{\mathrm{sym}}^{n\times n}$. Then the homogenized function

$$
f_{\mathrm{hom}}(X) := \inf_{k\in\mathbb{N}} \inf_{\varphi\in C_c^\infty(k\mathbb{I}^n;\mathbb{R}^n)} \frac{1}{k^n} \int_{k\mathbb{I}^n} f(x, X + \mathfrak{E}\varphi(x))\, dx
$$

is well-defined, satisfies the same growth condition, and for every open bounded convex set A and $\varepsilon_k \searrow 0$ it holds

$$
f_{\mathrm{hom}}(X) = \lim_{k\to\infty} \inf\left\{ \frac{1}{|\varepsilon_k^{-1}A|} \int_{\varepsilon_k^{-1}A} f(x, X + \mathfrak{E}\varphi(x))\, dx : \varphi \in C_c^\infty(\varepsilon_k^{-1}A; \mathbb{R}^n) \right\}.
$$

Moreover, f_{hom} is symmetric-quasiconvex and

$$
(f^{\mathrm{qcls}})_{\mathrm{hom}} = f_{\mathrm{hom}}.
$$

Proof. Most of the claims have been already deduced. For the upper bound we insert the zero function. The lower bound follows from convexity of the lower bound and Jensen's inequality. Define as above $f^{(\delta)}$, and choose any $\delta_j \searrow 0$. From Lemma (4.11) and Appendix A, it follows $\left((f^{(\delta_j)})^{\mathrm{qcls}}\right)_{\mathrm{hom}} = \left((f^{(\delta_j)})\right)_{\mathrm{hom}}$. Therefore, employing the monotonicity and referring to the dominated convergence theorem twice, we arrive at

$$(f^{\mathrm{qcls}})_{\mathrm{hom}}(X)$$

$$= \inf_{k\in\mathbb{N}} \inf_{\varphi\in C_c^\infty(k\mathbb{I}^n;\mathbb{R}^n)} \frac{1}{k^n} \int_{k\mathbb{I}^n} f^{\mathrm{qcls}}(x, X + \mathfrak{E}\varphi(x))\, dx$$

$$= \inf_{k\in\mathbb{N}} \inf_{\varphi\in C_c^\infty(k\mathbb{I}^n;\mathbb{R}^n)} \frac{1}{k^n} \int_{k\mathbb{I}^n} \inf_{\psi\in C_c^\infty(\mathbb{I}^n;\mathbb{R}^n)} \left(\int_{\mathbb{I}^n} f(x, X + \mathfrak{E}\varphi(x) + \mathfrak{E}\psi(y))\, dy \right) dx$$

$$= \inf_{k\in\mathbb{N}} \inf_{\varphi\in C_c^\infty(k\mathbb{I}^n;\mathbb{R}^n)} \frac{1}{k^n} \int_{k\mathbb{I}^n} \inf_{\psi\in C_c^\infty(\mathbb{I}^n;\mathbb{R}^n)} \left(\int_{\mathbb{I}^n} \inf_{j\in\mathbb{N}} f^{(\delta_j)}(x, X + \mathfrak{E}\varphi(x) + \mathfrak{E}\psi(y))\, dy \right) dx$$

$$= \inf_{k\in\mathbb{N}} \inf_{\varphi\in C_c^\infty(k\mathbb{I}^n;\mathbb{R}^n)} \frac{1}{k^n} \int_{k\mathbb{I}^n} \inf_{\psi\in C_c^\infty(\mathbb{I}^n;\mathbb{R}^n)} \left(\inf_{j\in\mathbb{N}} \int_{\mathbb{I}^n} f^{(\delta_j)}(x, X + \mathfrak{E}\varphi(x) + \mathfrak{E}\psi(y))\, dy \right) dx$$

$$= \inf_{k\in\mathbb{N}} \inf_{\varphi\in C_c^\infty(k\mathbb{I}^n;\mathbb{R}^n)} \frac{1}{k^n} \int_{k\mathbb{I}^n} \inf_{j\in\mathbb{N}} \inf_{\psi\in C_c^\infty(\mathbb{I}^n;\mathbb{R}^n)} \left(\int_{\mathbb{I}^n} f^{(\delta_j)}(x, X + \mathfrak{E}\varphi(x) + \mathfrak{E}\psi(y))\, dy \right) dx$$

$$= \inf_{k\in\mathbb{N}} \inf_{\varphi\in C_c^\infty(k\mathbb{I}^n;\mathbb{R}^n)} \inf_{j\in\mathbb{N}} \frac{1}{k^n} \int_{k\mathbb{I}^n} \inf_{\psi\in C_c^\infty(\mathbb{I}^n;\mathbb{R}^n)} \left(\int_{\mathbb{I}^n} f^{(\delta_j)}(x, X + \mathfrak{E}\varphi(x) + \mathfrak{E}\psi(y))\, dy \right) dx$$

$$= \inf_{j\in\mathbb{N}} \inf_{k\in\mathbb{N}} \inf_{\varphi\in C_c^\infty(k\mathbb{I}^n;\mathbb{R}^n)} \frac{1}{k^n} \int_{k\mathbb{I}^n} \inf_{\psi\in C_c^\infty(\mathbb{I}^n;\mathbb{R}^n)} \left(\int_{\mathbb{I}^n} f^{(\delta_j)}(x, X + \mathfrak{E}\varphi(x) + \mathfrak{E}\psi(y))\, dy \right) dx$$

$$= \inf_{j\in\mathbb{N}} \inf_{k\in\mathbb{N}} \inf_{\varphi\in C_c^\infty(k\mathbb{I}^n;\mathbb{R}^n)} \frac{1}{k^n} \int_{k\mathbb{I}^n} (f^{(\delta_j)})^{\mathrm{qcls}}(x, X + \mathfrak{E}\varphi(x))\, dx$$

$$= \inf_{j\in\mathbb{N}} \left((f^{(\delta_j)})^{\mathrm{qcls}} \right)_{\mathrm{hom}}(X)$$

$$= \inf_{j\in\mathbb{N}} (f^{(\delta_j)})_{\mathrm{hom}}(X)$$

$$= f_{\mathrm{hom}}(X). \quad \blacksquare$$

Let us apply this to our question. Define the functionals

$$\mathcal{G}(u) := \begin{cases} \int_\Omega f_{\mathrm{hom}}(\mathfrak{E}u(x))\, dx, & u \in LU(\Omega;\mathbb{R}^n), \\ \infty, & \text{else,} \end{cases}$$

and $\mathcal{G}^{(0)}$ by reducing the domain of $\mathcal{G}$ to $W^{1,2}(\Omega;\mathbb{R}^n)$. By the discussion above, for any $\delta_j \searrow 0$ we have the pointwise convergence $(f^{(\delta_j)})_{\mathrm{hom}} \searrow f_{\mathrm{hom}}$. It clearly follows

$$\mathcal{F}_{\mathrm{hom}}^{(\delta_j)}(u) \searrow \mathcal{G}^{(0)}(u)$$

for every $u \in L^1(\Omega;\mathbb{R}^n)$. Having a non-increasing sequence, it holds furthermore

$$(8.3) \qquad \Gamma(L^1)\text{-}\lim_{\delta\to 0} \mathcal{F}_{\mathrm{hom}}^{(\delta)} = \mathrm{lsc}\, \mathcal{G}^{(0)} = \mathrm{lsc}\, \mathcal{G}$$

where for the second equality we argue as in Subsection 7.2.1. From $\mathcal{F}_\varepsilon^{(\delta)} \geq \mathcal{F}_\varepsilon^{(0)}$ for every $\delta > 0$, it follows

$$\mathcal{F}_{\mathrm{hom}}^{(\delta)} \geq \Gamma(L^1)\text{-}\limsup_{\varepsilon\to 0} \mathcal{F}_\varepsilon^{(0)}.$$

The right-hand side is, being Γ-$\limsup$, lower semicontinuous. Hence, sending $\delta \to 0$ yields

$$\mathrm{lsc}\, \mathcal{G}^{(0)} \geq \Gamma(L^1)\text{-}\limsup_{\varepsilon\to 0} \mathcal{F}_\varepsilon^{(0)}.$$

Thus

$$(8.4) \qquad \mathrm{lsc}\, \mathcal{G} = \mathrm{lsc}\, \mathcal{G}^{(0)} \geq \Gamma(L^1)\text{-}\limsup_{\varepsilon\to 0} \mathcal{F}_\varepsilon^{(0)} \geq \Gamma(L^1)\text{-}\limsup_{\varepsilon\to 0} \mathcal{F}_\varepsilon.$$

8.3 $\langle\cdot\rangle$-strict continuity of $\mathcal{G}$

Let us first focus on the relaxation of $\mathcal{G}$. More precisely, we will look for a recovery sequence, i.e., try to determine $\Gamma(L^1)$-lim sup of the constant sequence $\mathcal{G}$.

We will make a more general analysis not solely restricted to our concrete problem. The denotations and assumptions will be determined in each step, and at the end we will return to our setting. We will study the $\langle\cdot\rangle$-strict continuity of functionals of form

$$U(\Omega;\mathbb{R}^n) \to \mathbb{R}, \quad u \mapsto \int_\Omega f\big(x, \mathfrak{E}u(x)\big) \, dx + \int_\Omega (f_{\mathrm{dev}})^\infty\big(x, \tfrac{dE^s u}{d|E^s u|}(x)\big) \, d|E^s u|(x).$$

The idea is to adapt Section 3 in [KR:10-1] to our purposes.

According to Theorem (7.11) and Remark (7.12), we may approximate every function in $U(\Omega;\mathbb{R}^n)$ with smooth functions in the $\langle\cdot\rangle$-strict topology. Therefore, the following form of the Reshetnyak continuity theorem (see Theorem 5 in [KR:10-1]) is applicable:

Theorem (8.5, Reshetnyak continuity theorem). *Let $f \in \mathbf{E}(\Omega;\mathbb{R}^N)$ and*

$$\mu_j \overset{*}{\rightharpoonup} \mu \quad in \ M(\Omega;\mathbb{R}^N) \quad and \quad \langle\mu_j\rangle(\Omega) \to \langle\mu\rangle(\Omega).$$

Then

$$\lim_{j\to\infty} \left[\int_\Omega f\left(x, \frac{d\mu_j^a}{d\mathcal{L}^n}(x)\right) \, dx + \int_\Omega f^\infty\left(x, \frac{d\mu_j^s}{d|\mu_j^s|}(x)\right) \, d|\mu_j^s|(x)\right] =$$
$$= \int_\Omega f\left(x, \frac{d\mu^a}{d\mathcal{L}^n}(x)\right) \, dx + \int_\Omega f^\infty\left(x, \frac{d\mu^s}{d|\mu^s|}(x)\right) \, d|\mu^s|(x).$$

Let us explain the denotations from the theorem. The recession function of some function $f : \Omega \times \mathbb{R}^N \to \mathbb{R}$ is defined as

$$f^\infty(x_0, X_0) := \limsup_{X \to X_0, \, t \to \infty} \frac{f(x_0, tX)}{t}.$$

A continuous function $f : \Omega \times \mathbb{R}^N \to \mathbb{R}$ belongs to $\mathbf{E}(\Omega;\mathbb{R}^N)$ if the function

$$Tf : \Omega \times B_1(0) \to \mathbb{R}, \quad (Tf)(x, \check{X}) := (1 - |\check{X}|)f(x, \tfrac{\check{X}}{1-|\check{X}|}),$$

has a bounded continuous extension to $\overline{\Omega \times B_1(0)}$. For these functions the recession function is actually the limit

$$f^\infty(x_0, X_0) = \lim_{\substack{x \to x_0, \\ X \to X_0, \\ t \to \infty}} \frac{f(x, tX)}{t}$$

and agrees on $\Omega \times \partial B_1(0)$ with the extension of Tf. With functions from $\mathbf{E}(\Omega;\mathbb{R}^N)$, we may approximate from below a large class of functions as was shown in Lemma 2.3 of [AB:97]:

Lemma (8.6). *Let* $f : \Omega \times \mathbb{R}^N \to \mathbb{R}$ *be lower semicontinuous such that for some* $\alpha > 0$ *it holds*

$$f(x, X) \geq -\alpha(1 + |X|)$$

for all $x \in \Omega$ *and* $X \in \mathbb{R}^n$. *There exists a non-decreasing sequence of functions* $\{g_k\}_{k \in \mathbb{N}}$ *from* $\mathbf{E}(\Omega; \mathbb{R}^n)$ *such that*

$$g_k(x, X) \geq -\alpha(1 + |X|), \quad \sup_{k \in \mathbb{N}} g_k = f \quad and \quad \sup_{k \in \mathbb{N}} g_k^\infty = h_f$$

where

$$h_f(x_0, X_0) := \liminf \left\{ \frac{f(x, tX)}{t} : x \to x_0, \ X \to X_0, \ t \to \infty \right\}.$$

Since h_f will play a significant role in the approximation of the recession function, let us give a more detailed formula for our setting.

Let $f : \Omega \times \mathbb{R}^{n \times n}_{\text{sym}} \to \mathbb{R}$. For $P_0 \in \mathbb{R}^{n \times n}_{\text{dev}}$ we may rewrite the definition in the following manner

$$
\begin{aligned}
h_f(x_0, P_0) &= \liminf \left\{ \frac{f(x, t(P + \frac{\rho}{n}I))}{t} : x \to x_0, \ P \to P_0 \text{ in } \mathbb{R}^{n \times n}_{\text{dev}}, \ \rho \to 0, \ t \to \infty \right\} \\
&= \sup_{k \in \mathbb{N}} \inf_{(x, P, \rho, t) \in E_{x_0, P_0, k}} \frac{f(x, t(P + \frac{\rho}{n}I))}{t}
\end{aligned}
$$

where

$$E_{x_0, P_0, k} := \left\{ (x, X, \rho, t) : |x - x_0| < \frac{1}{k}, \ P \in \mathbb{R}^{n \times n}_{\text{dev}}, |P - P_0| < \frac{1}{k}, |\rho| < \frac{1}{k}, t > k \right\}.$$

Notice that in $-h_{-f}$ lim inf is replaced by lim sup, i.e.,

$$
\begin{aligned}
-h_{-f}(x_0, P_0) &= \limsup \left\{ \frac{f(x, t(P + \frac{\rho}{n}I))}{t} : x \to x_0, \ P \to P_0 \text{ in } \mathbb{R}^{n \times n}_{\text{dev}}, \ \rho \to 0, \ t \to \infty \right\} \\
&= \inf_{k \in \mathbb{N}} \sup_{(x, P, \rho, t) \in E_{x_0, P_0, k}} \frac{f(x, t(P + \frac{\rho}{n}I))}{t}
\end{aligned}
$$

Our objective is to prove the following theorem.

Theorem (8.7). *Let* $f : \Omega \times \mathbb{R}^{n \times n}_{\text{sym}} \to \mathbb{R}$ *be a continuous function that is symmetric-rank-one-convex in the second variable and that satisfies the Hencky growth condition, i.e., there exist* $\alpha, \beta > 0$ *such that*

$$\alpha\big(|X_{\text{dev}}| + (\operatorname{tr} X)^2\big) \leq f(X, X) \leq \beta\big(1 + |X_{\text{dev}}| + (\operatorname{tr} X)^2\big).$$

Denote $f_{\text{dev}} := f|_{\Omega \times \mathbb{R}^{n \times n}_{\text{dev}}}$. *Suppose that*

$$(f_{\text{dev}})^\infty(x_0, P_0) = \limsup_{P \to P_0, t \to \infty} \frac{f_{\text{dev}}(x_0, tP)}{t}$$

is for every fixed $P_0 \in \mathbb{R}^{n \times n}_{\text{dev}}$ *a continuous function of* x_0. *Then the functional*

$$\mathcal{F}(u) = \int_\Omega f\big(x, \mathfrak{E}u(x)\big) \ dx + \int_\Omega (f_{\text{dev}})^\infty\big(x, \tfrac{dE^s u}{d|E^s u|}(x)\big) \ d|E^s u|(x)$$

is ⟨·⟩-*strictly continuous on* $U(\Omega; \mathbb{R}^n)$.

Remark (8.8).

(a) Since the deviatoric symmetric rank-one matrices span the whole $\mathbb{R}^{n\times n}_{\text{dev}}$, the function f_{dev} is globally Lipschitz in the second variable, i.e., there exists a constant C, depending only on n and β, such that

$$|f_{\text{dev}}(x, X) - f_{\text{dev}}(x, Y)| \le C|X - Y|$$

for all $x \in \Omega$ and $X, Y \in \mathbb{R}^{n\times n}_{\text{dev}}$. See Lemma (A.5).

(b) Consequently, the recession function is simply

$$(f_{\text{dev}})^{\infty}(x_0, P_0) = \limsup_{t\to\infty} \frac{f_{\text{dev}}(x_0, tP_0)}{t}.$$

For every $P_0 \in \mathbb{R}^{n\times n}_{\text{dev}}$ symmetric rank-one, this lim sup is even a limit (or a supremum) due to the convexity in the direction of P_0

$$
\begin{aligned}
\limsup_{t\to\infty} \frac{f_{\text{dev}}(x_0, tP_0)}{t} &= \lim_{t\to\infty} \frac{f_{\text{dev}}(x_0, tP_0) - f_{\text{dev}}(x_0, 0)}{t} \\
&= \sup_{t>0} \frac{f_{\text{dev}}(x_0, tP_0) - f_{\text{dev}}(x_0, 0)}{t}.
\end{aligned}
$$

(c) We will apply this theorem to an x-independent function. The continuity assumption on $(f_{\text{dev}})^{\infty}$ is in that case trivially fulfilled.

We wish to apply the Reshetnyak continuity theorem and Lemma (8.6). Therefore, we must carefully analyse the relationship between f^{∞} and $(f_{\text{dev}})^{\infty}$, and take into account the quadratic growth of f in the trace direction. For that reason, we prove a sort of Lipschitz continuity that enables us to compare values of finite and zero trace.

Lemma (8.9). *Let $f : \mathbb{R}^{n\times n}_{\text{sym}} \to \mathbb{R}$ be symmetric-rank-one-convex and suppose*

$$|f(X)| \le \beta(1 + |X_{\text{dev}}| + (\operatorname{tr} X)^2).$$

Then it fulfils the following local Lipschitz condition in the trace direction: For any $M, \varkappa \ge 1$ and $P \in \mathbb{R}^{n\times n}_{\text{dev}}$, it holds

$$|f(P + \tfrac{\rho}{n}I) - f(P)| \le 14\beta n\sqrt{\varkappa}(\sqrt{|P|} + M)|\rho|$$

for all $\rho^2 \le \varkappa(|P| + M^2)$.

The proof is based on Lemma (A.5). It says that for a separately convex function f on a ball $B_{2r}(X)$, its Lipschitz constant on $B_r(X)$ does not exceed $n\frac{\operatorname{osc}(f, B_{2r}(X))}{r}$. In comparison to the other standard result from Lemma (A.4), it is important that the assessment is local, and that it does not assume a global standard p-growth.

Proof. Let $P \in \mathbb{R}_{\mathrm{dev}}^{n \times n}$ be arbitrary. Then

$$
\begin{aligned}
\mathrm{osc}(f, B(P, 2r)) &= \sup_{X,Y \in B(P,2r)} |f(X) - f(Y)| \\
&\leq 2 \sup_{X \in B(P,2r)} |f(X)| \\
&\leq 2\beta(1 + |P| + 2r + 4r^2 n).
\end{aligned}
$$

Fix r by $r^2 n = \varkappa(|P| + M^2)$. Then

$$
\begin{aligned}
\tfrac{n}{r}\mathrm{osc}(f, B(P, 2r)) &\leq \tfrac{2n\beta}{r}(1 + r^2 n - M^2 + 2r + 4r^2 n) \\
&\leq 2n\beta(2 + 5rn) \\
&\leq 14n^{3/2}\beta\sqrt{\varkappa}(\sqrt{|P|} + M).
\end{aligned}
$$

The claim follows as $|\tfrac{\rho}{n}I| = \tfrac{\rho}{\sqrt{n}}$. ∎

Let f be as in Theorem (8.7). For any $M, K \in \mathbb{N}$ we set

$$
C_{M,K} := \{X \in \mathbb{R}_{\mathrm{sym}}^{n \times n} : |X_{\mathrm{dev}}| \geq K((\mathrm{tr}\, X)^2 - M^2)\}
$$

and define a lower and an upper bound for f

$$
\hat{f}_{M,K} \leq f \leq \check{f}_{M,K}
$$

in the following way. Let us choose a continuous function $\zeta_{M,K} : \mathbb{R}_{\mathrm{sym}}^{n \times n} \to [0,1]$ such that

$$
\zeta_{M,K}(X) = 1 \quad \text{for all } X \in C_{M,K} \quad \text{and} \quad \zeta_{M,K}(X) = 0 \quad \text{for every } X \notin C_{M+1,K}.
$$

Then define

$$
\hat{f}_{M,K}(x, X) := \zeta_{M,K}(X)f(x, X).
$$

The function $\hat{f}_{M,K}$ fulfils a linear growth condition since

$$
\hat{f}_{M,K}(x, X) \leq 1_{C_{M+1,K}}(X)\, f(x, X) \leq \beta(1 + 2|X_{\mathrm{dev}}| + (M + 1)^2).
$$

We define the upper bound as

$$
\check{f}_{M,K}(x, X) := f(x, X) + \beta K^2 \max\{(\mathrm{tr}\, X)^2 - M^2 - \tfrac{1}{K}|X_{\mathrm{dev}}|, 0\}.
$$

Lemma (8.10). *Suppose f is as in Theorem (8.7). For all $x_0 \in \Omega$, $P_0 \in \mathbb{R}_{\mathrm{dev}}^{n \times n}$, $M \geq 1$ and $K > 1$*

$$
-h_{-\hat{f}_{M,K}}(x_0, P_0) \leq -h_{-f_{\mathrm{dev}}}(x_0, P_0) + \frac{14\beta n}{\sqrt{K}}|P_0|
$$

and

$$
h_{\check{f}_{M,K}}(x_0, P_0) \geq h_{f_{\mathrm{dev}}}(x_0, P_0) - \frac{14\beta n}{\sqrt{K-1}}|P_0|.
$$

Proof. Let us fix $x_0 \in \Omega$ and $P_0 \in \mathbb{R}_{\mathrm{dev}}^{n \times n}$. We have to bound

$$
-h_{-\hat{f}_{M,K}}(x_0, P_0) = \inf_{k \in \mathbb{N}} \sup_{(x,P,\rho,t) \in E_{x_0,P_0,k}} \frac{\hat{f}_{M,K}(x, t(P + \tfrac{\rho}{n}I))}{t}.
$$

Take any $(x, P, \rho, t) \in E_{x_0, P_0, k}$.

- For $t(P + \frac{\rho}{n}I) \notin C_{M+1,K}$ we have $\hat{f}_{M,K}(x, t(P + \frac{\rho}{n}I)) = 0$.

- If $t(P + \frac{\rho}{n}I) \in C_{M+1,K}$, then $|t\rho| \leq \sqrt{\frac{1}{K}|tP| + (M+1)^2}$ and by Lemma (8.9)

$$
\begin{aligned}
\hat{f}_{M,K}(x, t(P + \tfrac{\rho}{n}I)) &\leq f(x, t(P + \tfrac{\rho}{n}I)) \\
&\leq f(x, tP) + 14\beta n(|tP|^{1/2} + M + 1)\sqrt{\tfrac{1}{K}|tP| + (M+1)^2}.
\end{aligned}
$$

In both cases

$$
\begin{aligned}
\frac{\hat{f}_{M,K}(x, t(P + \tfrac{\rho}{n}I))}{t} &\leq \frac{f(x, tP)}{t} + 14\beta n \left(\frac{M+1}{\sqrt{t}} + |P|^{1/2}\right)\sqrt{\frac{1}{K}|P| + \frac{(M+1)^2}{\sqrt{t}}} \\
&\leq \frac{f(x, tP)}{t} + 14\beta n \left(\frac{M+2}{\sqrt{k}} + |P_0|^{1/2}\right)\sqrt{\frac{1}{K}|P_0| + \frac{(M+2)^2}{\sqrt{k}}}.
\end{aligned}
$$

Hence

$$
-h_{-\hat{f}_{M,K}}(x_0, P_0) \leq -h_{-f_{\mathrm{dev}}}(x_0, P_0) + \frac{14\beta n}{\sqrt{K}}|P_0|.
$$

For

$$
h_{\check{f}_{M,K}}(x_0, P_0) = \sup_{k \in \mathbb{N}} \inf_{(x,P,\rho,t) \in E_{x_0,P_0,k}} \frac{\check{f}_{M,K}(x, t(P + \tfrac{\rho}{n}I))}{t}.
$$

again choose arbitrary $(x, P, \rho, t) \in E_{x_0, P_0, k}$.

- If $t(P + \frac{\rho}{n}I)) \notin C_{M+1,K-1}$, it holds $t^2\rho^2 \geq \frac{1}{K-1}t|P| + (M+1)^2$ and therefore

$$
\begin{aligned}
\check{f}_{M,K}(x, t(P + \tfrac{\rho}{n}I)) &\geq \beta K^2(t^2\rho^2 - \tfrac{1}{K}t|P| - M^2) \\
&\geq \beta K^2(\tfrac{1}{K-1}t|P| + (M+1)^2 - \tfrac{1}{K}t|P| - M^2) \\
&= \beta K^2(\tfrac{1}{K(K-1)}t|P| + 2M + 1) \\
&\geq \beta(t|P| + 1) \\
&\geq f(x, tP).
\end{aligned}
$$

- If $t(P + \frac{\rho}{n}I)) \in C_{M+1,K-1}$, we assess as above

$$
\begin{aligned}
\frac{\check{f}_{M,K}(x, t(P + \tfrac{\rho}{n}I))}{t} &\geq \frac{f(x, t(P + \tfrac{\rho}{n}I))}{t} \\
&\geq \frac{f(x, tP)}{t} - 14\beta n \left(\tfrac{M+1}{\sqrt{t}} + |P|^{1/2}\right)\sqrt{\tfrac{1}{K-1}|P| + \tfrac{(M+1)^2}{\sqrt{t}}} \\
&\geq \frac{f(x, tP)}{t} - 14\beta n \left(\tfrac{M+2}{\sqrt{k}} + |P_0|^{1/2}\right)\sqrt{\tfrac{1}{K-1}|P_0| + \tfrac{(M+2)^2}{\sqrt{k}}}.
\end{aligned}
$$

Therefore

$$
h_{\check{f}_{M,K}}(x_0, P_0) \geq h_{f_{\mathrm{dev}}}(x_0, P_0) - \frac{14\beta n}{\sqrt{K-1}}|P_0|. \quad \blacksquare
$$

In the following three lemmas, the assumptions of Theorem (8.7) should hold.

Lemma (8.11). *The functional* $\mathcal{F}^* : U(\Omega; \mathbb{R}^n) \to \mathbb{R}$

$$\mathcal{F}^*(u) := \int_\Omega f(x, \mathfrak{E}u(x))\ dx - \int_\Omega h_{-f_{\mathrm{dev}}}(x, \tfrac{dE^s u}{d|E^s u|}(x))\ d|E^s u|(x)$$

is upper semicontinuous with respect to the $\langle\cdot\rangle$*-strict topology.*

Proof. Let us take any sequence $u_j \xrightarrow{\langle\cdot\rangle} u$ in $U(\Omega; \mathbb{R}^n)$ and choose an arbitrary $K \in \mathbb{N}$. Since the sequence $\{(\operatorname{div} u_j)^2\}_{j\in\mathbb{N}}$ is equiintegrable, there exists $M \in \mathbb{N}$ such that

$$\int_{\{|\operatorname{div} u_j| > M\}} |\operatorname{div} u_j(x)|^2\ dx < \frac{1}{K^3}.$$

For every $j \in \mathbb{N}$ we split

$$\int_\Omega f(x, \mathfrak{E}u_j(x))\ dx = \int_{\{\mathfrak{E}u_j \in C_{M,K}\}} f(x, \mathfrak{E}u_j(x))\ dx + \int_{\{\mathfrak{E}u_j \notin C_{M,K}\}} f(x, \mathfrak{E}u_j(x))\ dx.$$

For the first term obviously

$$\int_{\{\mathfrak{E}u_j \in C_{M,K}\}} f(x, \mathfrak{E}u_j(x))\ dx = \int_{\{\mathfrak{E}u_j \in C_{M,K}\}} \hat{f}_{M,K}(x, \mathfrak{E}u_j(x))\ dx \le \int_\Omega \hat{f}_{M,K}(x, \mathfrak{E}u_j(x))\ dx.$$

By the definition $\mathfrak{E}u_j(x) \notin C_{M,K}$ means $|\mathfrak{E}_{\mathrm{dev}}u_j(x)| < K((\operatorname{div} u_j(x))^2 - M^2)$ and implies

$$|\operatorname{div} u_j(x)| > M \quad \text{and} \quad f(x, \mathfrak{E}u_j(x)) \le \beta(1 + 2K(\operatorname{div} u_j(x))^2 - KM^2)) \le 2K\beta(\operatorname{div} u_j(x))^2.$$

Therefore, for every $j \in \mathbb{N}$

$$\int_{\{\mathfrak{E}u_j \notin C_{M,K}\}} f(x, \mathfrak{E}u_j(x))\ dx \le \int_{\{|\operatorname{div} u_j| > M\}} 2K\beta(\operatorname{div} u_j(x))^2\ dx \le \frac{2\beta}{K^2}.$$

The singular part $E^s u$ is concentrated on $\mathbb{R}^{n\times n}_{\mathrm{dev}}$. Clearly $-h_{-f_{\mathrm{dev}}} \le -h_{-\hat{f}_{M,K}}$ (on deviatoric matrices). Hence, for

$$\hat{\mathcal{F}}_{M,K}(u) := \int_\Omega \hat{f}_{M,K}(x, \mathfrak{E}u(x))\ dx - \int_\Omega h_{-\hat{f}_{M,K}}(x, \tfrac{dE^s u}{d|E^s u|}(x))\ d|E^s u|(x)$$

we have $\mathcal{F}^*(u_j) \le \hat{\mathcal{F}}_{M,K}(u_j) + \frac{2\beta}{K^2}$. Since $\hat{f}_{M,K}$ grows linearly, we may approximate $-\hat{f}_{M,K}$ from below according to Lemma (8.6) with a sequence $\{g_k\}_{k\in\mathbb{N}} \subset \mathbf{E}(\Omega; \mathbb{R}^{n\times n}_{\mathrm{sym}})$. Hence, for every $k \in \mathbb{N}$

$$
\begin{aligned}
\liminf_{j\to\infty} -\hat{\mathcal{F}}_{M,K}(u_j) &\ge \liminf_{j\to\infty} \left(\int_\Omega g_k(x, \mathfrak{E}u_j(x))\ dx + \int_\Omega g_k^\infty(x, \tfrac{dE^s u_j}{d|E^s u_j|}(x))\ d|E^s u_j|(x) \right) \\
&\ge \int_\Omega g_k(x, \mathfrak{E}u(x))\ dx + \int_\Omega g_k^\infty(x, \tfrac{dE^s u}{d|E^s u|}(x))\ d|E^s u|(x)
\end{aligned}
$$

since for g_k we may apply the Reshetnyak continuity theorem. Hence, by the monotone convergence theorem

$$\liminf_{j\to\infty} -\hat{\mathcal{F}}_{M,K}(u_j) \ge -\hat{\mathcal{F}}_{M,K}(u).$$

By gathering the results above we get

$$\limsup_{j\to\infty} \mathcal{F}^*(u_j) \leq \limsup_{j\to\infty} \hat{\mathcal{F}}_{M,K}(u_j) + \frac{2\beta}{K^2} \leq \hat{\mathcal{F}}_{M,K}(u) + \frac{2\beta}{K^2}.$$

By Lemma (8.10)

$$
\begin{aligned}
\hat{\mathcal{F}}_{M,K}(u) &= \int_\Omega \hat{f}_{M,K}(x, \mathfrak{E}u(x))\, dx - \int_\Omega h_{-\hat{f}_{M,K}}\left(x, \tfrac{dE^s u}{d|E^s u|}(x)\right)\, d|E^s u|(x) \\
&\leq \int_\Omega f(x, \mathfrak{E}u(x))\, dx - \int_\Omega h_{-f_{\mathrm{dev}}}\left(x, \tfrac{dE^s u}{d|E^s u|}(x)\right)\, d|E^s u|(x) + \frac{14\beta n}{\sqrt{K}}|E^s u|(\Omega).
\end{aligned}
$$

Altogether,

$$\limsup_{j\to\infty} \mathcal{F}^*(u_j) \leq \mathcal{F}^*(u) + \frac{2\beta}{K^2} + \frac{14\beta n}{\sqrt{K}}|E^s u|(\Omega).$$

Since K was arbitrary, the upper semicontinuity follows. ∎

Lemma (8.12). *The functional* $\mathcal{F}_* : U(\Omega; \mathbb{R}^n) \to \mathbb{R}$

$$\mathcal{F}_*(u) := \int_\Omega f(x, \mathfrak{E}u(x))\, dx + \int_\Omega h_{f_{\mathrm{dev}}}\left(x, \tfrac{dE^s u}{d|E^s u|}(x)\right)\, d|E^s u|(x)$$

is lower semicontinuous with respect to the $\langle \cdot \rangle$-*strict topology.*

Proof. The proof is very similar to the previous one. Let us therefore just point out the differences. Here we employ $\check{f}_{M,K}$, which yields the lower bound

$$
\begin{aligned}
\int_\Omega f(x, \mathfrak{E}u_j(x))\, dx &\geq \int_\Omega \check{f}_{M,K}(x, \mathfrak{E}u_j(x))\, dx - \beta K^2 \int_{\{\mathfrak{E}u_j \notin C_{M,K}\}} (\operatorname{div} u_j)^2\, dx \\
&\geq \int_\Omega \check{f}_{M,K}(x, \mathfrak{E}u_j(x))\, dx - \frac{\beta}{K}.
\end{aligned}
$$

Since on deviatoric matrices $h_{f_{\mathrm{dev}}} \geq h_{\check{f}_{M,K}}$, we arrive at $\mathcal{F}_*(u_j) \geq \check{\mathcal{F}}_{M,K}(u_j) - \frac{\beta}{K}$ where

$$\check{\mathcal{F}}_{M,K}(u) := \int_\Omega \check{f}_{M,K}(x, \mathfrak{E}u(x))\, dx + \int_\Omega h_{\check{f}_{M,K}}\left(x, \tfrac{dE^s u}{d|E^s u|}(x)\right)\, d|E^s u|(x).$$

The function $\check{f}_{M,K}$ meets the assumptions of Lemma (8.6). By the same argumentation and by Lemma (8.10) we get

$$\liminf_{j\to\infty} \mathcal{F}_*(u_j) \geq \mathcal{F}_*(u) - \frac{\beta}{K} - \frac{14\beta n}{\sqrt{K-1}}|E^s u|(\Omega). \quad ∎$$

Lemma (8.13). *For every* $x_0 \in \Omega$ *and symmetric rank-one matrix* $P_0 \in \mathbb{R}_{\mathrm{dev}}^{n\times n}$

$$(f_{\mathrm{dev}})^\infty(x_0, P_0) = h_{f_{\mathrm{dev}}}(x_0, P_0) = -h_{-f_{\mathrm{dev}}}(x_0, P_0).$$

The proof is exactly the same as in Lemma 1 in [KR:10-1].

Proof. Fix any $x_0 \in \Omega$ and any symmetric rank-one matrix $P_0 \in \mathbb{R}^{n \times n}_{\text{dev}}$. Then

$$\frac{f_{\text{dev}}(x, tP)}{t} = \frac{f_{\text{dev}}(x, tP) - f_{\text{dev}}(x, tP_0)}{t} + \frac{f_{\text{dev}}(x, tP_0) - f_{\text{dev}}(x, 0)}{t} + \frac{f_{\text{dev}}(x, 0)}{t}$$

By Remark (8.8) we have

$$\frac{|f_{\text{dev}}(x, tP) - f_{\text{dev}}(x, tP_0)|}{t} \leq \frac{C|tP - tP_0|}{t} = C|P - P_0|,$$

and we know that the functions

$$g_t(x) := \frac{f_{\text{dev}}(x, tP_0) - f_{\text{dev}}(x, 0)}{t}$$

make a monotonically increasing family with

$$g_t(x) \nearrow (f_{\text{dev}})^\infty(x, P_0)$$

for every $x \in \Omega$. All g_t are continuous, and for $(f_{\text{dev}})^\infty(\_, P_0)$ continuity was an assumption (see Theorem (8.7)). Therefore, we may apply Dini's Lemma. Hence,

$$g_t \nearrow (f_{\text{dev}})^\infty(\_, P_0)$$

uniformly on $\{x \in \Omega : |x - x_0| \leq \frac{1}{k}\}$ for every $k \in \mathbb{N}$. Clearly we have also $|\frac{f_{\text{dev}}(x,0)}{t}| \leq \frac{\beta}{t}$. Gathering all the estimates yields

$$\begin{aligned} h_{f_{\text{dev}}}(x_0, P_0) &= \liminf \left\{ \frac{f_{\text{dev}}(x, tP)}{t} : x \to x_0, \ P \to P_0, \ t \to \infty \right\} \\ &= \liminf_{x \to x_0} (f_{\text{dev}})^\infty(x, P_0) \\ &= (f_{\text{dev}})^\infty(x_0, P_0) \end{aligned}$$

as well as

$$-h_{-f_{\text{dev}}}(x_0, P_0) = \limsup_{x \to x_0} (f_{\text{dev}})^\infty(x, P_0) = (f_{\text{dev}})^\infty(x_0, P_0). \quad \blacksquare$$

Proof (of Theorem (8.7)). According to Theorem (7.3), for $|E^s u|$-a.e. $x \in \Omega$, the Radon-Nikodym derivative $\frac{dE^s u}{d|E^s u|}(x)$ is a symmetric-rank-one matrix. By Lemma (8.13) the functionals from Lemmas (8.11) and (8.12) coincide with $\mathcal{F}$. $\quad \blacksquare$

Gathering the density result from Theorem (7.11) and the continuity result in Theorem (8.7), we get:

Corollary (8.14). *Let $f : \mathbb{R}^{n \times n}_{\text{sym}} \to \mathbb{R}$ be a symmetric-rank-one-convex function that satisfies a Hencky plasticity growth condition. For every $u \in U(\Omega; \mathbb{R}^n)$ there exists a sequence $\{u_j\}_{j \in \mathbb{N}} \subset C^\infty(\Omega; \mathbb{R}^n) \cap LU(\Omega; \mathbb{R}^n)$ with $\gamma(u_j) = \gamma(u)$ such that*

- $u_j \xrightarrow{\langle \cdot \rangle} u$ *in* $U(\Omega; \mathbb{R}^n)$,

- $\displaystyle \lim_{j \to \infty} \int_\Omega f(\mathfrak{E}u_j(x)) \ dx = \int_\Omega f(\mathfrak{E}u(x)) \ dx + \int_\Omega (f_{\text{dev}})^\infty \left(\frac{dE^s u}{d|E^s u|}(x) \right) d|E^s u|(x).$

Now, let us return to our setting. We have shown that $\mathcal{G}|_{LU(\Omega;\mathbb{R}^n)}$, its density being symmetric-quasiconvex, is continuous in the $\langle\cdot\rangle$-strict topology. Moreover, its continuous extension to $U(\Omega;\mathbb{R}^n)$ in this topology is given by

$$\overline{\mathcal{G}}(u) := \int_\Omega f_{\text{hom}}(\mathfrak{E}u(x))\ dx + \int_\Omega \left((f_{\text{hom}})_{\text{dev}}\right)^\infty \left(\tfrac{dE^su}{d|E^su|}(x)\right)\ d|E^su|(x).$$

Thus,

(8.15)
$$(\text{lsc }\mathcal{G})|_{U(\Omega;\mathbb{R}^n)} \leq \overline{\mathcal{G}}.$$

8.4   lim inf-inequality at zero hardening

Now we turn our attention to the lim inf-inequality. Let us take any $u \in L^1(\Omega;\mathbb{R}^n)$, and choose $\varepsilon_j \searrow 0$ and $u_j \to u$ in $L^1(\Omega;\mathbb{R}^n)$. Clearly, if $\liminf_{j\to\infty} \mathcal{F}_{\varepsilon_j}(u_j) = \infty$, there is nothing to be proved. If

$$\liminf_{j\to\infty} \mathcal{F}_{\varepsilon_j}(u_j) < \infty,$$

there exists a subsequence $\{j_k\}_{k\in\mathbb{N}}$ such that

$$\liminf_{j\to\infty} \mathcal{F}_{\varepsilon_j}(u_j) = \lim_{k\to\infty} \mathcal{F}_{\varepsilon_{j_k}}(u_{j_k})$$

with all elements being finite. Hence, $\{u_{j_k}\}_{k\in\mathbb{N}}$ is bounded in $LU(\Omega;\mathbb{R}^n)$, and there exists a further (not relabelled) sequence that weakly converges in $U(\Omega;\mathbb{R}^n)$ (see Subsection 7.2.1). Therefore, $u \in U(\Omega;\mathbb{R}^n)$. By Theorem (7.2), $u_{j_k} \to u$ in $L^q(\Omega;\mathbb{R}^n)$ for all $1 < q < \frac{n}{n-1}$. Moreover, we may also achieve that the measures

$$\mu_k := f(\tfrac{\cdot}{\varepsilon_{j_k}}, \mathfrak{E}u_{j_k}(\cdot))\ \mathcal{L}^n$$

weakly-$*$ converge to some μ in $M(\Omega;\mathbb{R}^n)$. Let

$$\mu = g\mathcal{L}^n + \mu^s$$

be the decomposition according to the Radon-Nikodym theorem. Our aim will be to determine the derivative g and the singular part μ^s, as

(8.16)
$$\liminf_{j\to\infty} \mathcal{F}_{\varepsilon_j}(u_j) = \lim_{k\to\infty} \mu_k(\Omega) \geq \mu(\Omega) = \int_\Omega g(x)\ dx + \mu^s(\Omega).$$

The discussion above has shown that we may restrict ourselves to the following setting: We consider arbitrary $u \in U(\Omega;\mathbb{R}^n)$, $\varepsilon_j \searrow 0$ and a bounded sequence $\{u_j\}_{j\in\mathbb{N}} \subset LU(\Omega;\mathbb{R}^n)$ such that

- $\lim_{j\to\infty} \mathcal{F}_{\varepsilon_j}(u_j)$ exists,

- $u_j \rightharpoonup u$ in $U(\Omega;\mathbb{R}^n)$ and $u_j \to u$ in $L^q(\Omega;\mathbb{R}^n)$ for a fixed $1 < q < \frac{n}{n-1}$,

- $f(\tfrac{\cdot}{\varepsilon_j}, \mathfrak{E}u_j(\cdot))\ \mathcal{L}^n =: \mu_j \overset{*}{\rightharpoonup} \mu =: g\mathcal{L}^n + \mu^s$ in $M(\Omega;\mathbb{R}^n)$.

We will bound g from below in Lemma (8.17) and, under an additional assumption, μ^s in Lemma (8.21).

8.4.1 Regular points

Lemma (8.17). *For a.e. $x_0 \in \Omega$, it holds*

$$g(x_0) \geq f_{\mathrm{hom}}(\mathfrak{C}u(x_0)).$$

We will follow the strategy of the proofs of Propositions 11.2.3 and 12.3.2 in [ABM:06].

Proof. By the Besicovitch derivation theorem (C.1), for a.e. $x_0 \in \Omega$ we have

(8.18)
$$g(x_0) = \lim_{\rho \to 0} \frac{\mu(B_\rho(x_0))}{|B_\rho(x_0)|}.$$

According to Lemmas (C.2) and (C.3), for all but countable many $\rho > 0$ it holds

$$\mu(B_\rho(x_0)) = \lim_{j \to \infty} \mu_j(B_\rho(x_0)).$$

Therefore, it must be shown for a.e. $x_0 \in \Omega$

$$\lim_{\rho \to 0} \lim_{j \to \infty} \frac{\mu_j(B_\rho(x_0))}{|B_\rho(x_0)|} \geq f_{\mathrm{hom}}(\mathfrak{C}u(x_0))$$

(whereby we exclude the exceptional sequence of ρ's).

Let us take and fix any x_0 where the formula (8.18) holds and where the function u is L^q-differentiable (see Corollary (7.6)), and define

$$\tilde{u}(x) := u(x_0) + \nabla u(x_0) \, (x - x_0).$$

We may also suppose $\mathfrak{C}\tilde{u} = \mathfrak{C}\tilde{u}(x_0) = \operatorname{sym} \nabla u(x_0)$. Our strategy is to approximate u with $\tilde{u}$ and to use the slicing method of De Giorgi. Therefore, choose any $\nu \in \mathbb{N}$ and $0 < \lambda < 1$ and define

$$\rho_0 := \lambda \rho \qquad \text{and} \qquad B_i := B_{\rho_0 + \frac{i}{\nu}(\rho - \rho_0)}(x_0), \quad i = 0, \ldots, \nu.$$

Furthermore, we take for every $i = 1, \ldots, \nu$ also cut-off functions $\varphi_i \in C_c^\infty(B_i)$ such that

$$0 \leq \varphi_i \leq 1, \quad \varphi_i = 1 \text{ in } B_{i-1} \quad \text{and} \quad \|\nabla \varphi_i\|_{L^\infty} \leq \frac{2\nu}{\rho - \rho_0}.$$

Let

$$\tilde{u}_{j,i} := \tilde{u} + \varphi_i(u_j - \tilde{u}) \in L^1(\Omega; \mathbb{R}^n).$$

Because of

$$\mathfrak{C}\tilde{u}_{j,i} = (1 - \varphi_i)\mathfrak{C}\tilde{u} + \varphi_i \mathfrak{C}u_j + \nabla \varphi_i \odot (u_j - \tilde{u}),$$

the functions $\tilde{u}_{j,i}$ lie in $LD(\Omega; \mathbb{R}^n)$, but perhaps not in $LU(\Omega; \mathbb{R}^n)$ since

$$\operatorname{div} \tilde{u}_{j,i} = (1 - \varphi_i) \operatorname{div} \tilde{u} + \varphi_i \operatorname{div} u_j + \nabla \varphi_i \cdot (u_j - \tilde{u}),$$

and the last term in general lies only in $L^{\frac{n}{n-1}}$. We will correct this with the results of Bogovskii from Theorem (7.17). For that reason define

$$\zeta_{j,i} := \frac{1}{|B_i \setminus B_{i-1}|} \int_{B_i \setminus B_{i-1}} \nabla \varphi_i(x) \cdot (u_j(x) - \tilde{u}(x)) \, dx.$$

By Theorem (7.17) there exist $z_{j,i} \in W_0^{1,q}(B_i \setminus \overline{B_{i-1}})$ such that

$$\operatorname{div} z_{j,i} = -\nabla \varphi_i \cdot (u_j - \tilde{u}) + \zeta_{j,i}$$

with

$$\|z_{j,i}\|_{W^{1,q}(B_i \setminus \overline{B_{i-1}})} \leq C \|\nabla \varphi_i \cdot (u_j - \tilde{u})\|_{L^q(B_i \setminus \overline{B_{i-1}})} \leq \frac{2C\nu}{\rho - \rho_0} \|u_j - \tilde{u}\|_{L^q(B_i \setminus \overline{B_{i-1}})}.$$

We take such constant C that the inequality holds for all i. It is scaling and translation invariant, so we may transfer the situation to $B_1(0)$. Therefore, C does not depend on ρ. Although it depends on ν and λ, this will not cause any troubles since we will first send $j \to \infty$. Now define $u_{j,i} := \tilde{u}_{j,i} + z_{j,i} \in LU(\Omega; \mathbb{R}^n)$. By the discussion in Section 7.3,

$$u_{j,i} - \tilde{u} = \varphi_i(u_j - \tilde{u}) + z_{j,i} \in LU_0(B_\rho(x_0); \mathbb{R}^n).$$

For every $i = 1, \ldots, \nu$

$$
\begin{aligned}
&f_{\text{hom}}(\mathfrak{E}u(x_0)) \\
&= \liminf_{j \to \infty} \left\{ \frac{1}{|\frac{1}{\varepsilon_j} B_\rho(x_0)|} \int_{\frac{1}{\varepsilon_j} B_\rho(x_0)} f(x, \mathfrak{E}u(x_0) + \mathfrak{E}\varphi(x)) \, dx : \varphi \in LU_0(\tfrac{1}{\varepsilon_j} B_\rho(x_0), \mathbb{R}^n) \right\} \\
&= \liminf_{j \to \infty} \left\{ \frac{1}{|B_\rho(x_0)|} \int_{B_\rho(x_0)} f(\tfrac{x}{\varepsilon_j}, \mathfrak{E}u(x_0) + \mathfrak{E}\varphi(x)) \, dx : \varphi \in LU_0(B_\rho(x_0), \mathbb{R}^n) \right\} \\
&\leq \liminf_{j \to \infty} \frac{1}{|B_\rho(x_0)|} \int_{B_\rho(x_0)} f(\tfrac{x}{\varepsilon_j}, \mathfrak{E}u_{j,i}(x)) \, dx,
\end{aligned}
$$

and therefore also

$$f_{\text{hom}}(\mathfrak{E}u(x_0)) \leq \liminf_{j \to \infty} \frac{1}{\nu} \sum_{i=1}^{\nu} \frac{1}{|B_\rho(x_0)|} \int_{B_\rho(x_0)} f(\tfrac{x}{\varepsilon_j}, \mathfrak{E}u_{j,i}(x)) \, dx.$$

For every i we split

$$
\begin{aligned}
&\int_{B_\rho(x_0)} f(\tfrac{x}{\varepsilon_j}, \mathfrak{E}u_{j,i}(x)) \, dx \\
&= \int_{B_{i-1}} f(\tfrac{x}{\varepsilon_j}, \mathfrak{E}u_j(x)) \, dx + \int_{B_i \setminus B_{i-1}} f(\tfrac{x}{\varepsilon_j}, \mathfrak{E}u_{j,i}(x)) \, dx + \int_{B_\rho(x_0) \setminus B_i} f(\tfrac{x}{\varepsilon_j}, \mathfrak{E}u(x_0)) \, dx \\
&=: I_{j,i}^{(1)} + I_{j,i}^{(2)} + I_{j,i}^{(3)}.
\end{aligned}
$$

The first term can be bounded simply by

$$I_{j,i}^{(1)} \leq \int_{B_\rho(x_0)} f(\tfrac{x}{\varepsilon_j}, \mathfrak{E}u_j(x)) \, dx$$

and the last one by

$$I_{j,i}^{(3)} \leq \beta n(1 - \lambda)|B_\rho(x_0)| \big(1 + |\mathfrak{E}_{\text{dev}}u(x_0)| + |\operatorname{div} u(x_0)|^2\big).$$

The second term we bound by the upper bound on f

$$f\left(\tfrac{x}{\varepsilon_j}, \mathfrak{E}u_{j,i}(x)\right) \le \beta(1 + |\mathfrak{E}_{\mathrm{dev}}u_{j,i}(x)| + (\mathrm{div}\, u_{j,i}(x))^2).$$

First,

$$
\begin{aligned}
|\mathfrak{E}_{\mathrm{dev}}u_{j,i}| &= |(1-\varphi_i)\mathfrak{E}_{\mathrm{dev}}\tilde{u} + \varphi_i\mathfrak{E}_{\mathrm{dev}}u_j + \mathrm{dev}\,\mathrm{sym}(\nabla\varphi_i \otimes (u_j - \tilde{u})) + \mathfrak{E}_{\mathrm{dev}}z_{j,i}| \\
&\le |\mathfrak{E}_{\mathrm{dev}}\tilde{u}| + |\mathfrak{E}_{\mathrm{dev}}u_j| + |\nabla\varphi_i||u_j - \tilde{u}| + |\mathfrak{E}_{\mathrm{dev}}z_{j,i}|.
\end{aligned}
$$

First we bound the last two terms

$$\fint_{B_i\setminus B_{i-1}} |\nabla\varphi_i(x)||u_j(x) - \tilde{u}(x)|\, dx \le \frac{2\nu|B_i\setminus B_{i-1}|^{1/q'}}{(1-\lambda)\rho}\left(\fint_{B_i\setminus B_{i-1}} |u_j(x) - \tilde{u}(x)|^q\, dx\right)^{1/q}$$

and

$$
\begin{aligned}
\fint_{B_i\setminus B_{i-1}} |\mathfrak{E}_{\mathrm{dev}}z_{j,i}(x)|\, dx &\le |B_i\setminus B_{i-1}|^{1/q'}\left(\fint_{B_i\setminus B_{i-1}} |\mathfrak{E}_{\mathrm{dev}}z_{j,i}(x)|^q\, dx\right)^{1/q} \\
&\le \frac{2C\nu|B_i\setminus B_{i-1}|^{1/q'}}{(1-\lambda)\rho}\left(\fint_{B_i\setminus B_{i-1}} |u_j(x) - \tilde{u}(x)|^q\, dx\right)^{1/q}.
\end{aligned}
$$

(q' stands for the Hölder conjugate of q.) Thus, by applying $|B_i\setminus B_{i-1}| \le \frac{n(1-\lambda)}{\nu}|B_\rho(x_0)|$, we arrive at

$$
\begin{aligned}
\fint_{B_i\setminus B_{i-1}} &|\mathfrak{E}_{\mathrm{dev}}u_{j,i}(x)|\, dx \\
&\le |B_i\setminus B_{i-1}||\mathfrak{E}_{\mathrm{dev}}u(x_0)| + \fint_{B_i\setminus B_{i-1}} |\mathfrak{E}_{\mathrm{dev}}u_j(x)|\, dx + \\
&\quad + \frac{2(1+C)\nu^{1/q}}{(1-\lambda)^{1/q}}n^{1/q'}|B_\rho(x_0)|^{1/q'}\left(\fint_{B_i\setminus B_{i-1}} \frac{|u_j(x) - \tilde{u}(x)|^q}{\rho^q}\, dx\right)^{1/q}.
\end{aligned}
$$

Also

$$(\mathrm{div}\, u_{j,i})^2 = \left((1-\varphi_i)\,\mathrm{div}\,\tilde{u} + \varphi_i\,\mathrm{div}\, u_j + \zeta_{j,i}\right)^2 \le 3(\mathrm{div}\,\tilde{u})^2 + 3(\mathrm{div}\, u_j)^2 + 3\zeta_{j,i}^2,$$

and therefore,

$$
\begin{aligned}
\fint_{B_i\setminus B_{i-1}} (\mathrm{div}\, u_{j,i}(x))^2\, dx &\le 3|B_i\setminus B_{i-1}|(\mathrm{div}\, u(x_0))^2 + \\
&\quad + 3\fint_{B_i\setminus B_{i-1}} (\mathrm{div}\, u_j(x))^2\, dx + 3\frac{n(1-\lambda)}{\nu}|B_\rho(x_0)|\zeta_{j,i}^2.
\end{aligned}
$$

For a bound for the last term, we proceed as above

$$
\begin{aligned}
|\zeta_{j,i}| &\le \frac{1}{|B_i\setminus B_{i-1}|}\fint_{B_i\setminus B_{i-1}} |\nabla\varphi_i(x)||u_j(x) - \tilde{u}(x)|\, dx \\
&\le \frac{2\nu}{(1-\lambda)|B_i\setminus B_{i-1}|^{1/q}}\left(\fint_{B_i\setminus B_{i-1}} \frac{|u_j(x) - \tilde{u}(x)|^q}{\rho^q}\, dx\right)^{1/q} \\
&\le \frac{2\nu^{1+1/q}}{n^{1/q}\lambda^{n/q}(1-\lambda)^{1+1/q}}\left(\frac{1}{|B_\rho(x_0)|}\fint_{B_i\setminus B_{i-1}} \frac{|u_j(x) - \tilde{u}(x)|^q}{\rho^q}\, dx\right)^{1/q},
\end{aligned}
$$

where we used $|B_i \setminus B_{i-1}| \geq \frac{n\lambda^n(1-\lambda)}{\nu}|B_\rho(x_0)|$. Then

$$\sum_{i=1}^{\nu} |\zeta_{j,i}| \leq \frac{\nu^2}{n^{1/q}\lambda^{n/q}(1-\lambda)^{1+1/q}} \left(\frac{1}{|B_\rho(x_0)|}\int_{B_\rho(x_0)} \frac{|u_j(x)-\tilde{u}(x)|^q}{\rho^q}\,dx\right)^{1/q}.$$

Altogether for the second term

$$
\begin{aligned}
I_{j,i}^{(2)} \leq\ & 3\beta\Bigg[|B_i\setminus B_{i-1}|\big(1+|\mathfrak{E}_{\mathrm{dev}}u(x_0)|+(\operatorname{div}u(x_0))^2\big)+ \\
& +\int_{B_i\setminus B_{i-1}} \big(|\mathfrak{E}_{\mathrm{dev}}u_j(x)|+(\operatorname{div}u_j(x))^2\big)\,dx+ \\
& +\frac{2(1+C)\nu}{(1-\lambda)^{1/q}}n^{1/q'}|B_\rho(x_0)|\left(\frac{1}{|B_\rho(x_0)|}\int_{B_i\setminus B_{i-1}}\frac{|u_j(x)-\tilde{u}(x)|^q}{\rho^q}\,dx\right)^{1/q}+ \\
& +\frac{n(1-\lambda)}{\nu}|B_\rho(x_0)|\zeta_{j,i}^2\Bigg]
\end{aligned}
$$

and

$$\sum_{i=1}^{\nu}I_{j,i}^{(2)}$$

$$
\begin{aligned}
\leq\ & 3\beta\Bigg[n(1-\lambda)|B_\rho(x_0)|\big(1+|\mathfrak{E}_{\mathrm{dev}}u(x_0)|+|\operatorname{div}u(x_0)|^2\big)+ \\
& +\int_{B_\rho(x_0)}\frac{f\big(\frac{x}{\varepsilon_j},\mathfrak{E}u_j(x)\big)}{\alpha}\,dx+ \\
& +\frac{2(1+C)\nu}{(1-\lambda)^{1/q}}(n\nu)^{1/q'}|B_\rho(x_0)|\left(\frac{1}{|B_\rho(x_0)|}\int_{B_\rho(x_0)}\frac{|u_j(x)-\tilde{u}(x)|^q}{\rho^q}\,dx\right)^{1/q}+ \\
& +\frac{n(1-\lambda)}{\nu}|B_\rho(x_0)|\frac{\nu^4}{n^{2/q}\lambda^{2n/q}(1-\lambda)^{2+2/q}}\left(\frac{1}{|B_\rho(x_0)|}\int_{B_\rho(x_0)}\frac{|u_j(x)-\tilde{u}(x)|^q}{\rho^q}\,dx\right)^{2/q}\Bigg].
\end{aligned}
$$

Now

$$
\begin{aligned}
& \frac{1}{\nu}\sum_{i=1}^{\nu}\frac{I_{j,i}^{(1)}+I_{j,i}^{(2)}+I_{j,i}^{(3)}}{|B_\rho(x_0)|} \\
\leq\ & \frac{1}{|B_\rho(x_0)|}\int_{B_\rho(x_0)}\Big(1+\tfrac{3\beta}{\alpha\nu}\Big)f\big(\tfrac{x}{\varepsilon_j},\mathfrak{E}u_j(x)\big)\,dx+ \\
& +4\beta n(1-\lambda)\big(1+|\mathfrak{E}_{\mathrm{dev}}u(x_0)|+|\operatorname{div}u(x_0)|^2\big)+ \\
& +\frac{2(1+C)}{(1-\lambda)^{1/q}}(n\nu)^{1/q'}\left(\frac{1}{|B_\rho(x_0)|}\int_{B_\rho(x_0)}\frac{|u_j(x)-\tilde{u}(x)|^q}{\rho^q}\,dx\right)^{1/q}+ \\
& +\frac{\nu^2 n^{1-2/q}}{\lambda^{2n/q}(1-\lambda)^{1+2/q}}\left(\frac{1}{|B_\rho(x_0)|}\int_{B_\rho(x_0)}\frac{|u_j(x)-\tilde{u}(x)|^q}{\rho^q}\,dx\right)^{2/q}.
\end{aligned}
$$

We now send $j \to \infty$ and use in the last two terms that $u_j \to u$ in $L^q(\Omega; \mathbb{R}^n)$. Hence,

$$
\begin{aligned}
f_{\text{hom}}(\mathfrak{E}u(x_0)) \ &\leq \ \liminf_{j \to \infty} \frac{1}{\nu} \sum_{i=1}^{\nu} \frac{I_{j,i}^{(1)} + I_{j,i}^{(2)} + I_{j,i}^{(3)}}{|B_\rho(x_0)|} \\
&\leq \ \left(1 + \tfrac{3\beta}{\alpha\nu}\right) \lim_{j \to \infty} \frac{\mu_j(B_\rho(x_0))}{|B_\rho(x_0)|} + \\
&\quad + 4\beta n(1 - \lambda)\left(1 + |\mathfrak{E}_{\text{dev}}u(x_0)| + |\operatorname{div} u(x_0)|^2\right) \\
&\quad + \frac{2(1 + C)}{(1 - \lambda)^{1/q}} (n\nu)^{1/q'} \left(\frac{1}{|B_\rho(x_0)|} \int_{B_\rho(x_0)} \frac{|u(x) - \tilde{u}(x)|^q}{\rho^q}\, dx\right)^{1/q} + \\
&\quad + \frac{\nu^2 n^{1-2/q}}{\lambda^{2n/q}(1 - \lambda)^{1+2/q}} \left(\frac{1}{|B_\rho(x_0)|} \int_{B_\rho(x_0)} \frac{|u(x) - \tilde{u}(x)|^q}{\rho^q}\, dx\right)^{2/q}.
\end{aligned}
$$

Sending also $\rho \to 0$ (excluding countably many) and applying L^q-differentiability of u yields

$$
\begin{aligned}
f_{\text{hom}}(\mathfrak{E}u(x_0)) \ &\leq \ \left(1 + \tfrac{3\beta}{\alpha\nu}\right) \lim_{\rho \to 0} \lim_{j \to \infty} \frac{\mu_j(B_\rho(x_0))}{|B_\rho(x_0)|} + \\
&\quad + 4\beta n(1 - \lambda)\left(1 + |\mathfrak{E}_{\text{dev}}u(x_0)| + |\operatorname{div} u(x_0)|^2\right).
\end{aligned}
$$

Since $\lambda < 1$ and $\nu \in \mathbb{N}$ were arbitrary, we get

$$
f_{\text{hom}}(\mathfrak{E}u(x_0)) \leq \lim_{\rho \to 0} \lim_{j \to \infty} \frac{\mu_j(B_\rho(x_0))}{|B_\rho(x_0)|}. \quad \blacksquare
$$

Now we may state a partial result on the Γ-limit.

Corollary (8.19). *If $u \in U(\Omega; \mathbb{R}^n)$ and $u_j \to u$ in $L^1(\Omega; \mathbb{R}^n)$, then*

$$
\liminf_{j \to \infty} \mathcal{F}_{\varepsilon_j}(u_j) \geq \int_\Omega f_{\text{hom}}(\mathfrak{E}u(x))\, dx
$$

for every $\varepsilon_j \searrow 0$. Therefore, for $u \in LU(\Omega; \mathbb{R}^n)$

$$
\Gamma(L^1)\text{-}\lim_{\varepsilon \to 0} \mathcal{F}_\varepsilon(u) = \int_\Omega f_{\text{hom}}(\mathfrak{E}u(x))\, dx.
$$

Proof. Only the second statement needs some arguing. Gathering the conclusions (8.4) and (8.15) (with the same denotations) yields

$$
\Gamma(L^1)\text{-}\limsup_{\varepsilon \to 0} \mathcal{F}_\varepsilon(u) \leq \operatorname{lsc} \mathcal{G}(u) \leq \overline{\mathcal{G}}(u) = \int_\Omega f_{\text{hom}}(\mathfrak{E}u(x))\, dx
$$

for every $u \in LU(\Omega; \mathbb{R}^n)$. $\quad \blacksquare$

8.4.2   Singular points

Until now, apart from the general assumptions on periodicity and growth, we have not imposed any kind of convexity on the density f. In order to control the behaviour in the singular points, we will have to assume f to be governed by a convex function at infinity. More precisely, we suppose that

- there is a Carathéodory function $c : \mathbb{R}^n \times \mathbb{R}^{n\times n}_{\mathrm{sym}} \to \mathbb{R}$ that is $\mathbb{I}^n$-periodic in the first variable and convex in the second,

- for every $\eta > 0$ there is $\beta_\eta > 0$

such that for a.e. $x \in \mathbb{R}^n$ and all $X \in \mathbb{R}^{n\times n}_{\mathrm{sym}}$

$$|f(x, X) - c(x, X)| \le \eta|X| + \beta_\eta.$$

We will refer to this property as *asymptotic convexity*.

Let us notice that for f in our setting we may even suppose c to be non-negative with $c(x, 0) = 0$ for every $x \in \mathbb{R}^n$. Namely, if $\tilde{c}$ is some function with the above properties (with coefficients $\tilde{\beta}_\eta$), then we may take $c(x, X) := \max\{\tilde{c}(x, X) - \beta - \tilde{\beta}_1, \alpha(|X_{\mathrm{dev}}| + (\mathrm{tr}\, X)^2)\}$. Then $\beta_\eta := \tilde{\beta}_\eta + \beta + \tilde{\beta}_1$. Moreover, c also has a Hencky plasticity growth and may be chosen with the coefficients $\hat{\alpha} := \alpha$ and $\hat{\beta} := \beta + 1 + \beta_1$. The motivation for these assumptions are the results from [DQ:90] for convex densities. Let us present them here more precisely.

Suppose that $c : \mathbb{R}^n \times \mathbb{R}^{n\times n}_{\mathrm{sym}} \to \mathbb{R}$ is a Carathéodory function that

- is $\mathbb{I}^n$-periodic in the first variable and convex in the second,

- for some $\hat{\alpha}, \hat{\beta} > 0$ fulfils

$$\hat{\alpha}(|X_{\mathrm{dev}}| + (\mathrm{tr}\, X)^2) \le c(x, X) \le \hat{\beta}(|X_{\mathrm{dev}}| + (\mathrm{tr}\, X)^2 + 1)$$

  for all $x \in \mathbb{R}^n$ and $X \in \mathbb{R}^{n\times n}_{\mathrm{sym}}$,

- is non-negative with $c(x, 0) = 0$ for all $x \in \mathbb{R}^n$.

We introduce

$$\mathcal{C}_\varepsilon(u) := \begin{cases} \int_\Omega c\left(\frac{x}{\varepsilon}, \mathfrak{E}u(x)\right) dx, & u \in LU(\Omega; \mathbb{R}^n), \\ \infty, & \text{else.} \end{cases}$$

In [DQ:90] the authors introduce for every non-negative function $\varphi \in C(\overline{\Omega})$ also the functionals

$$\langle \mathcal{C}_\varepsilon(u), \varphi \rangle := \begin{cases} \int_\Omega c\left(\frac{x}{\varepsilon}, \mathfrak{E}u(x)\right) \varphi(x)\, dx, & u \in LU(\Omega; \mathbb{R}^n), \\ \infty, & \text{else.} \end{cases}$$

In Theorem 1.1 they show that for any $\varepsilon_j \searrow 0$ there exists a subsequence $\{j_k\}_{k\in\mathbb{N}}$ such that $\Gamma(L^q)\text{-}\lim_{k\to\infty}\langle \mathcal{C}_{\varepsilon_{j_k}}(u), \varphi \rangle$ exists for every non-negative continuous $\varphi : \overline{\Omega} \to \mathbb{R}$ and every $u \in U(\Omega; \mathbb{R}^n)$. ($q$ is as before, i.e. $1 < q < \frac{n}{n-1}$.) In Proposition 2.1 it is proved that for $u \in LU(\Omega; \mathbb{R}^n)$ the corresponding Γ-limit is given by a density, which is by Proposition 2.2 location-independent.

If $\varphi = 1$, we get the existence of $\Gamma(L^1)$-$\lim_{k\to\infty} \mathcal{C}_{\varepsilon_{j_k}}$ on the whole $L^1(\Omega; \mathbb{R}^n)$ with the domain $U(\Omega; \mathbb{R}^n)$. The passage from the $\Gamma(L^q)$-limit to the $\Gamma(L^1)$-limit follows from the lower bound and the compactness of the embedding $U(\Omega; \mathbb{R}^n) \hookrightarrow L^q(\Omega; \mathbb{R}^n)$. By Corollary (8.19) the density of the Γ-limit must be c_{hom}.

Morever, they show that this density determines the Γ-limit for every $u \in U(\Omega; \mathbb{R}^n)$ with the formula

$$\int_\Omega \varphi(x) \, d\big(c_{\text{hom}}(Eu)\big)(x).$$

Under the integral there is a measure $c_{\text{hom}}(Eu)$ that still needs to be explained. Before that, let us notice that the expression neither depends on $\{\varepsilon_j\}_{j\in\mathbb{N}}$ nor on $\{j_k\}_{k\in\mathbb{N}}$. By the Urysohn property it follows that actually even $\Gamma(L^q)$-$\lim_{\varepsilon\to 0}\langle\mathcal{C}_\varepsilon(\_), \varphi\rangle$ and therefore $\Gamma(L^1)$-$\lim_{\varepsilon\to 0} \mathcal{C}_\varepsilon$ exist and are given by c_{hom}.

Now we return to the definition of $c_{\text{hom}}(Eu)$. For convex functions with a possible super-linear growth, this was done in [DT:86]. However, there are some requirements that have to be met (see Subsection 2.2 therein). Right away we see that c_{hom} is a non-negative finite convex function with $c_{\text{hom}}(0) = 0$. We will denote the *asymptotic function* of c_{hom} by

$$(c_{\text{hom}})^\#(X) := \lim_{t\to\infty} \frac{c_{\text{hom}}(tX)}{t}.$$

Since c_{hom} has superlinear growth, the definition does not coincide with the recession function. The latter is in this case for all $X \in \mathbb{R}^{n\times n}_{\text{sym}}$

$$(c_{\text{hom}})^\infty(X) = \limsup_{t\to\infty, \, Y\to X} \frac{c_{\text{hom}}(tY)}{t} = \infty.$$

This distinction is actually a very important issue in this analysis. To emphasize the difference, our denotation differs from the one in [DT:86]. The domain of $(c_{\text{hom}})^\#$ is $\mathbb{R}^{n\times n}_{\text{dev}}$, and for $X \in \mathbb{R}^{n\times n}_{\text{dev}}$ it holds $c_{\text{hom}}(X) \leq \beta(|X| + 1)$. According to Proposition 1.2 from the same source, another assumption is thus fulfilled. Let us also notice that

$$(c_{\text{hom}})^\#\big|_{\mathbb{R}^{n\times n}_{\text{dev}}} = (c_{\text{hom}}\big|_{\mathbb{R}^{n\times n}_{\text{dev}}})^\infty,$$

see Remark (8.8).

The last requirement concerns the conjugate function

$$(c_{\text{hom}})^*(Y) = \sup_{X\in\mathbb{R}^{n\times n}_{\text{sym}}} \big(X \cdot Y - c_{\text{hom}}(X)\big).$$

It is supposed that

(8.20)   domain of $(c_{\text{hom}})^*$ is closed.

Since this does not follow from the previous assumptions, we additionally assume this property. Then, according to Section 2.2 in [DT:86], we may define

$$c_{\text{hom}}(Eu) := c_{\text{hom}}(\mathfrak{E}u) \, \mathcal{L}^n + (c_{\text{hom}})^\#\big(\tfrac{dE^s u}{d|E^s u|}\big) \, |E^s u|.$$

Therefore, for every non-negative $\varphi \in C(\overline{\Omega})$ and $u \in U(\Omega; \mathbb{R}^n)$,

$$\Gamma(L^q)\text{- } \lim_{\varepsilon \to \infty} \langle \mathcal{C}_\varepsilon(u), \varphi \rangle = \langle \mathcal{C}_{\text{hom}}(u), \varphi \rangle := \int_\Omega \varphi(x) \, dc_{\text{hom}}(Eu)(x).$$

As before this implies

$$\Gamma(L^1)\text{- } \lim_{\varepsilon \to \infty} \mathcal{C}_\varepsilon = \mathcal{C}_{\text{hom}}$$

where

$$\mathcal{C}_{\text{hom}}(u) := \begin{cases} \int_\Omega c_{\text{hom}}(Eu), & u \in U(\Omega; \mathbb{R}^n), \\ \infty, & \text{else.} \end{cases}$$

To simplify the denotation, let us define the asymptotic function also for non-convex functions as

$$f^\#(X) := \limsup_{t \to \infty} \frac{f(tX)}{t}.$$

As already stated in Remark (8.8), if f is Lipschitz continuous, it coincides with the recession function.

Lemma (8.21). *Let f from our setting additionally be asymptotically convex, and let the corresponding convex function c fulfil (8.20). Then*

$$\mu^s \geq (f_{\text{hom}})^\# \left(\frac{dE^s u}{d|E^s u|} \right) |E^s u|.$$

Proof. Take any non-negative $\varphi \in C_0(\Omega)$. We may add to our assumptions on $\{u_j\}_{j \in \mathbb{N}}$ that $|\mathfrak{E}u_j| \mathcal{L}^n \overset{*}{\rightharpoonup} \sigma$ in $M(\Omega)$. Because of the weak-$*$ convergence and from

$$f(x, X) \geq c(x, X) - \eta|X| - \beta_\eta,$$

it follows

$$
\begin{aligned}
\int_\Omega \varphi(x) \, d\mu(x) &= \lim_{j \to \infty} \int_\Omega \varphi(x) \, d\mu_j(x) \\
&\geq \liminf_{j \to \infty} \langle \mathcal{C}_{\varepsilon_j}(u_j), \varphi \rangle - \lim_{j \to \infty} \eta \int_\Omega |\mathfrak{E}u_j(x)| \varphi(x) \, dx - \beta_\eta \int_\Omega \varphi(x) \, dx \\
&\geq \langle \mathcal{C}_{\text{hom}}(u), \varphi \rangle - \eta \langle \sigma, \varphi \rangle - \beta_\eta \int_\Omega \varphi(x) \, dx.
\end{aligned}
$$

Hence, $\mu \geq c_{\text{hom}}(Eu) - \eta \sigma - \beta_\eta \mathcal{L}^n$. The inequality holds also for the corresponding singular part, i.e.,

$$\mu^s \geq (c_{\text{hom}})^\# \left(\frac{dE^s u}{d|E^s u|} \right) |E^s u| - \eta \sigma^s.$$

Since this holds for any $\eta > 0$, even

$$\mu^s \geq (c_{\text{hom}})^\# \left(\frac{dE^s u}{d|E^s u|} \right) |E^s u|.$$

From

$$|X| \leq |X_{\text{dev}}| + (\operatorname{tr} X)^2 + 1 \leq \frac{c(x, X)}{\hat{\alpha}} + 1,$$

it follows

$$\left(1 - \frac{\eta}{\hat{\alpha}} \right) c(x, X) - \eta - \beta_\eta \leq f(x, X) \leq \left(1 + \frac{\eta}{\hat{\alpha}} \right) c(x, X) + \eta + \beta_\eta.$$

Hence,

$$\left(1 - \frac{\eta}{\hat{\alpha}}\right) c_{\text{hom}}(X) - \eta - \beta_\eta \leq f_{\text{hom}}(X) \leq \left(1 + \frac{\eta}{\hat{\alpha}}\right) c_{\text{hom}}(X) + \eta + \beta_\eta.$$

Therefore,

$$\left(1 - \frac{\eta}{\hat{\alpha}}\right) (c_{\text{hom}})^{\#}(X) \leq (f_{\text{hom}})^{\#}(X) \leq \left(1 + \frac{\eta}{\hat{\alpha}}\right) (c_{\text{hom}})^{\#}(X).$$

This holds for every $\eta > 0$. Therefore, $(f_{\text{hom}})^{\#} = (c_{\text{hom}})^{\#}$, and

$$\mu^s \geq (f_{\text{hom}})^{\#}\left(\frac{dE^s u}{d|E^s u|}\right)|E^s u|. \quad \blacksquare$$

8.5   Conclusion

Let us here gather the results in a compact way. For the homogenization at zero hardening, we have shown the following:

Theorem (8.22). *Let us have a Carathéodory function $f : \mathbb{R}^n \times \mathbb{R}^{n \times n}_{\text{sym}} \to \mathbb{R}$ that is*

- *$\mathbb{I}^n$-periodic in the first variable,*

- *there exist $\alpha, \beta > 0$ such that for a.e. $x \in \Omega$ and every $X \in \mathbb{R}^{n \times n}_{\text{sym}}$*

$$\alpha(|X_{\text{dev}}| + (\text{tr } X)^2) \leq f(x, X) \leq \beta(|X_{\text{dev}}| + (\text{tr } X)^2 + 1).$$

Let us denote

$$\mathcal{F}_\varepsilon(u) := \begin{cases} \int_\Omega f\left(\frac{x}{\varepsilon}, \mathfrak{E}u(x)\right) \, dx, & u \in LU(\Omega; \mathbb{R}^n), \\ \infty, & \text{else,} \end{cases}$$

and

$$\mathcal{F}_{\text{hom}}(u) := \begin{cases} \int_\Omega f_{\text{hom}}(\mathfrak{E}u(x)) \, dx + \int_\Omega (f_{\text{hom}})^{\#}\left(\frac{dE^s u}{d|E^s u|}(x)\right) \, d|E^s u|(x), & u \in U(\Omega; \mathbb{R}^n), \\ \infty, & \text{else.} \end{cases}$$

Then

$$\Gamma(L^1)\text{-}\limsup_{\varepsilon \to 0} \mathcal{F}_\varepsilon \leq \mathcal{F}_{\text{hom}},$$

while for $u \in LU(\Omega; \mathbb{R}^n)$ even

$$\Gamma(L^1)\text{-}\lim_{\varepsilon \to 0} \mathcal{F}_\varepsilon(u) = \mathcal{F}_{\text{hom}}(u).$$

The latter holds for all $u \in L^1(\Omega; \mathbb{R}^n)$ if f is asymptotically convex and if for the corresponding convex function (8.20) holds.

Proof. By (8.4)

$$\Gamma(L^1)\text{-}\limsup_{\varepsilon \to 0} \mathcal{F}_\varepsilon \leq \text{lsc } \mathcal{G},$$

where

$$\mathcal{G}(u) = \begin{cases} \int_\Omega f_{\text{hom}}(\mathfrak{E}u(x)) \, dx, & u \in LU(\Omega; \mathbb{R}^n), \\ \infty, & \text{else.} \end{cases}$$

By Remark (8.8), it holds $\left((f_{\text{hom}})_{\text{dev}}\right)^\infty = (f_{\text{hom}})^{\#}|_{\mathbb{R}^{n \times n}_{\text{dev}}}$. Therefore, by (8.15)

$$\text{lsc } \mathcal{G} \leq \mathcal{F}_{\text{hom}}.$$

The lim inf-inequality follows from (8.16) with Lemmas (8.17) and (8.21). $\blacksquare$

Now we give an answer also to the commutability of homogenization and vanishing hardening:

Theorem (8.23). *We suppose that the assumptions of Theorem (8.22), including the asymptotic convexity, hold. With $\mathcal{F}_\varepsilon^{(\delta)}$ and $\mathcal{F}_{\text{hom}}^{(\delta)}$ for $\delta \geq 0$ as in Section 8.1, the following diagrams commute:*

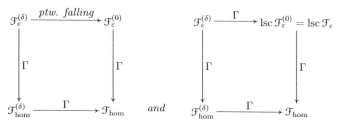

All Γ-limits are with respect to the L^1-norm.

Proof. The upper and the left-hand arrows were derived in Section 8.1 and depicted in (8.1). Both right-hand arrows follow from Theorem (8.22) and the properties of Γ-convergence (see Appendix B). Finally, (8.3), (8.4) and (8.15) imply

$$\Gamma(L^1)\text{-}\lim_{\delta \to 0} \mathcal{F}_{\text{hom}}^{(\delta)} = \mathcal{F}_{\text{hom}}. \quad \blacksquare$$

8.6   Relaxation at zero hardening revisited

The considerations of the previous sections may be applied to the functionals with location-independent densities. Clearly, thus we are investigating the relaxation of the homogeneous setting with or without hardening. In this case, however, we know in advance that the Γ-limit exists. The result for this simplified case at zero hardening reads as follows.

Corollary (8.24). *Let us have a continuous function $f : \mathbb{R}^{n \times n}_{\text{sym}} \to \mathbb{R}$ with a Hencky plasticity growth. The lower semicontinuous envelope (in $L^1(\Omega; \mathbb{R}^n)$) of the functional*

$$\mathcal{F}(u) := \begin{cases} \int_\Omega f\big(\mathfrak{E}u(x)\big) \ dx, & u \in LU(\Omega; \mathbb{R}^n), \\ \infty, & \text{else,} \end{cases}$$

is for $u \in LU(\Omega; \mathbb{R}^n)$ given by

$$\text{lsc} \, \mathcal{F}(u) = \int_\Omega f^{\text{qcls}}\big(\mathfrak{E}u(x)\big) \ dx$$

and is bounded from above for every $u \in U(\Omega; \mathbb{R}^n)$ by

$$\text{lsc} \, \mathcal{F}(u) \leq \int_\Omega f^{\text{qcls}}\big(\mathfrak{E}u(x)\big) \ dx + \int_\Omega (f^{\text{qcls}})^{\#}\Big(\tfrac{dE^s u}{d|E^s u|}(x)\Big) \ d|E^s u|(x).$$

If f is asymptotically convex (with (8.20)), then this inequality is even an equality for all $u \in U(\Omega; \mathbb{R}^n)$.

We would like to reconsider the assumption of asymptotic convexity for this case taking into account the recent progress in this field, i.e. the results in [KK:16]. For the sake of simplicity, we start with f that is already symmetric-quasiconvex.

Suppose that f for some $0 \leq \gamma < 1$ fulfils

$$f(X) \geq f^\#(X_{\mathrm{dev}}) - \beta(|X_{\mathrm{dev}}|^\gamma + 1)$$

for all $X \in \mathbb{R}^{n \times n}_{\mathrm{sym}}$. Here we actually have two assumptions in mind.

First, we make the projection

$$f(X) \geq f(X_{\mathrm{dev}}) - C_1(|X_{\mathrm{dev}}|^\gamma + 1).$$

Recalling the quadratic growth in the trace direction, it is not a very strong assumption. However, it cannot be excluded just by using the lower bound and Lemma (8.9). Currently, it is not known to us whether alone the symmetric-quasiconvexity with a Hencky plasticity growth condition suffices for this estimate.

Additionally, we suppose for $P \in \mathbb{R}^{n \times n}_{\mathrm{dev}}$ also

$$f(P) \geq f^\#(P) - C_2(|P|^\gamma + 1).$$

We met such a behaviour already in Chapter 5, and it is in accordance with similar assumptions in the literature (e.g., [BFT:00, BFM:98]).

Thus, we imposed a lower bound on the function f, however, not by a convex function as before, but by a 1-homogeneous symmetric-rank-one-convex function (on $\mathbb{R}^{n \times n}_{\mathrm{dev}}$). Let us be more precise:

- $V := \mathbb{R}^{n \times n}_{\mathrm{dev}}$ is a finite-dimensional normed space,

- $D := \{a \odot b : a, b \in \mathbb{R}^n, \ a \perp b\}$ spans V,

- $g := f^\#|_{\mathbb{R}^{n \times n}_{\mathrm{dev}}} : V \to \mathbb{R}$ is 1-homogeneous and is convex along any direction from D.

According to Theorem 1.1 in [KK:16], g is convex at every point from D. To be more specific, for every pair of orthogonal vectors $a, b \in \mathbb{R}^n$, there exists a (homogeneous) linear function $\ell : V \to \mathbb{R}$ such that

$$g \geq \ell \quad \text{everywhere on } V \quad \text{and} \quad g(a \odot b) = \ell(a \odot b).$$

With this tools we may prove the alternative version of the corollary above.

Proposition (8.25). *Let us have a symmetric-quasiconvex function* $f : \mathbb{R}^{n \times n}_{\mathrm{sym}} \to \mathbb{R}$ *for which there exist* $\alpha, \beta > 0$ *and* $\gamma \in [0, 1)$ *such that*

$$\alpha(|X_{\mathrm{dev}}| + (\operatorname{tr} X)^2) \leq f(X) \leq \beta(|X_{\mathrm{dev}}| + (\operatorname{tr} X)^2 + 1)$$

and

$$f(X) \geq f^\#(X_{\mathrm{dev}}) - \beta(|X_{\mathrm{dev}}|^\gamma + 1)$$

for all $X \in \mathbb{R}^{n \times n}_{\text{sym}}$. *Then the lower semicontinuous envelope of the functional*

$$\mathcal{F}(u) := \begin{cases} \int_\Omega f\big(\mathfrak{E}u(x)\big) \ dx, & u \in LU(\Omega; \mathbb{R}^n), \\ \infty, & \text{else,} \end{cases}$$

is

$$\text{lsc } \mathcal{F}(u) = \begin{cases} \int_\Omega f\big(\mathfrak{E}u(x)\big) \ dx + \int_\Omega f^\#\big(\frac{dE^s u}{d|E^s u|}(x)\big) \ d|E^s u|(x), & u \in U(\Omega; \mathbb{R}^n), \\ \infty, & \text{else.} \end{cases}$$

Proof. With Corollary (8.14) and Lemma (8.17) still being valid, we just have to prove the lim inf-inequality in the singular points.

Let us take any $u \in U(\Omega; \mathbb{R}^n)$ and let $u_j \to u$ in $L^1(\Omega; \mathbb{R}^n)$. Clearly, we may for $\{u_j\}_{j \in \mathbb{N}}$ consider only bounded sequences in $LU(\Omega; \mathbb{R}^n)$. Moreover, we may suppose

- $u_j \rightharpoonup u$ in $U(\Omega; \mathbb{R}^n)$,

- $\mu_j := f(\mathfrak{E}u_j)\mathcal{L}^n$ converge weakly-$*$ to some μ in $M(\Omega; \mathbb{R}^n)$,

- $|\mathfrak{E}_{\text{dev}}u_j|^\gamma + 1$ converge weakly to some h in $L^{1/\gamma}(\Omega; \mathbb{R}^n)$.

Our goal is to show $\mu^s \geq f^\#(\frac{dE^s u}{d|E^s u|})|E^s u|$ with $\mu = g\,\mathcal{L}^n + \mu^s$ again being the Lebesgue decomposition of μ with respect to $\mathcal{L}^n$.

According to Theorem (7.3), for $|E^s u|$-a.e. $x_0 \in \Omega$ there exist $a(x_0), b(x_0) \in \mathbb{R}^n$ such that

$$(8.26) \qquad \frac{dE^s u}{d|E^s u|}(x_0) = \lim_{\rho \to 0} \frac{Eu(B_\rho(x_0))}{|Eu|(B_\rho(x_0))} = a(x_0) \odot b(x_0).$$

Since $\text{tr } E^s u = 0$, we have $\text{tr}\,(a(x_0) \odot b(x_0)) = a(x_0) \cdot b(x_0) = 0$. By the Besicovitch derivation theorem (C.1), for $|E^s u|$-a.e. $x_0 \in \Omega$ also

$$(8.27) \qquad \frac{d\mu^s}{d|E^s u|}(x_0) = \lim_{\rho \to 0} \frac{\mu(B_\rho(x_0))}{|E^s u|(B_\rho(x_0))} = \lim_{\rho \to 0} \frac{\mu(B_\rho(x_0))}{|Eu|(B_\rho(x_0))}.$$

The second equality follows from the fact that

$$|Eu| = |\mathfrak{E}u|\mathcal{L}^n + |E^s u| \quad \text{and} \quad |\mathfrak{E}u|\mathcal{L}^n \perp |E^s u|.$$

Hence, for $|E^s u|$-a.e. $x_0 \in \Omega$

$$\lim_{\rho \to 0} \frac{\int_{B_\rho(x_0)} |\mathfrak{E}u(x)| \ dx}{|E^s u|(B_\rho(x_0))} = 0.$$

Since $(\text{div } u)\,\mathcal{L}^n$ and $h\,\mathcal{L}^n$ are each mutually singular with $|E^s u|$, for $|E^s u|$-a.e. $x_0 \in \Omega$ also

$$(8.28) \qquad \lim_{\rho \to 0} \frac{\int_{B_\rho(x_0)} \text{div } u(x) \ dx}{|Eu|(B_\rho(x_0))} = \lim_{\rho \to 0} \frac{\int_{B_\rho(x_0)} h(x) \ dx}{|Eu|(B_\rho(x_0))} = 0.$$

Let $x_0 \in \Omega$ be from now on any point where (8.26), (8.27) and (8.28) hold. Being fixed, we stop writing x_0 in the denotations. We need to show that

$$\lim_{\rho \to 0} \frac{\mu(B_\rho)}{|Eu|(B_\rho)} \geq f^\#(a \odot b).$$

Let $\ell : \mathbb{R}^{n \times n}_{\mathrm{dev}} \to \mathbb{R}$ be a linear function from Theorem 1.1 in [KK:16] that determines the supporting hyperplane for $f^\#|_{\mathbb{R}^{n \times n}_{\mathrm{dev}}}$ at $a \odot b$. According to Lemmas (C.2) and (C.3), for all but countable many $\rho > 0$ it holds

$$
\begin{aligned}
\mu(B_\rho) &= \lim_{j \to \infty} \mu_j(B_\rho) \\
&= \lim_{j \to \infty} \int_{B_\rho} f(\mathfrak{E}u_j(x)) \, dx \\
&\geq \limsup_{j \to \infty} \int_{B_\rho} f^\#(\mathfrak{E}_{\mathrm{dev}}u_j(x)) \, dx - \beta \lim_{j \to \infty} \int_{B_\rho} \left(|\mathfrak{E}_{\mathrm{dev}}u_j(x)|^\gamma + 1 \right) dx \\
&\geq \limsup_{j \to \infty} \int_{B_\rho} \ell(\mathfrak{E}_{\mathrm{dev}}u_j(x)) \, dx - \beta \int_{B_\rho} h(x) \, dx \\
&= \ell\big(\mathrm{dev}\, Eu(B_\rho) \big) - \beta \int_{B_\rho} h(x) \, dx.
\end{aligned}
$$

In the last equality above, we used the linearity of ℓ and applied Lemmas (C.2) and (C.3) to $(\mathfrak{E}u_j)\mathcal{L}^n \overset{*}{\rightharpoonup} Eu$. Hence, by (8.28) and (8.26)

$$\lim_{\rho \to 0} \frac{\mu(B_\rho)}{|Eu|(B_\rho)} \geq \limsup_{\rho \to 0} \ell\left(\frac{\mathrm{dev}\, Eu(B_\rho)}{|Eu|(B_\rho)} \right) = \ell\left(\lim_{\rho \to 0} \frac{Eu(B_\rho)}{|Eu|(B_\rho)} \right) = \ell(a \odot b).$$

Now, we employ

$$\ell(a \odot b) = f^\#(a \odot b). \qquad \blacksquare$$

Appendix

Appendix A

Different types of convexity

Although standard in the calculus of variations, we repeat some convexity properties of functions.

Definition (A.1). A locally bounded Borel function $f : \mathbb{R}^{m \times n} \to \mathbb{R}$ is

- *quasiconvex* if for every $X \in \mathbb{R}^{m \times n}$ and every $\varphi \in C_c^\infty(\mathbb{I}^n; \mathbb{R}^m)$

$$\int_{\mathbb{I}^n} f(X + \nabla\varphi(x)) \, dx \geq f(X),$$

- *rank-one convex* if for every $X \in \mathbb{R}^{m \times n}$ and every pair $a \in \mathbb{R}^m$, $b \in \mathbb{R}^n$ the function

$$t \mapsto f(X + t \, a \otimes b)$$

 is convex,

- *separately convex* if for every $X \in \mathbb{R}^{m \times n}$ and every $i \in \{1, \ldots, m\}$ and $j \in \{1, \ldots, n\}$ the function

$$t \mapsto f(X + t \, e_i \otimes e_j)$$

 is convex.

Every type of convexity in the previous definition implies the following.

By f^{qc} we denote the *quasiconvex envelope* of f, i.e., the largest quasiconvex function that does not exceed f. If f is bounded from below at least by some quasiconvex function, it may be computed by

$$(A.2) \qquad f^{\mathrm{qc}}(X) = \inf_{\varphi \in W_0^{1,\infty}(U;\mathbb{R}^m)} \frac{1}{|U|} \int_U f(X + \nabla\varphi(x)) \, dx$$

where $U \subset \mathbb{R}^n$ is arbitrary bounded open set.

For functions of the form $f : \Omega \times \mathbb{R}^{m \times n} \to \mathbb{R}$, the envelope f^{qc} is to be computed with the first argument being fixed. A thorough coverage of this topic can be found, e.g., in [Dac:08].

There are also corresponding definitions on the space of symmetric matrices.

Definition (A.3). A locally bounded Borel function $f : \mathbb{R}_{\text{sym}}^{n \times n} \to \mathbb{R}$ is *symmetric-quasi-convex* (resp. *symmetric-rank-one convex*) if the function

$$\mathbb{R}^{n \times n} \to \mathbb{R}, \quad X \mapsto f(X_{\text{sym}})$$

is quasiconvex (resp. rank-one convex).

Therefore, a symmetric-quasiconvex function f must fulfil

$$\int_{\mathbb{I}^n} f(X + \mathfrak{E}\varphi(x)) \, dx \geq f(X)$$

for every $X \in \mathbb{R}_{\text{sym}}^{n \times n}$ and every $\varphi \in C_c^\infty(\mathbb{I}^n; \mathbb{R}^n)$ whereas symmetric-rank-one convexity means that

$$t \mapsto f(X + t\, a \odot b)$$

is convex for all $X \in \mathbb{R}_{\text{sym}}^{n \times n}$ and $a, b \in \mathbb{R}^n$. We denote the *symmetric-quasiconvex envelope* by f^{qcls}. It is related to the quasiconvex envelope by the formula

$$(f \circ \text{sym})^{\text{qc}} = f^{\text{qcls}} \circ \text{sym}.$$

(sym : $\mathbb{R}^{n \times n} \to \mathbb{R}_{\text{sym}}^{n \times n}$ is simply the symmetrizing projection.) For the proof and other properties, we refer to [Zh:04].

A function with any of these convexity properties and with a suitable upper bound is locally Lipschitz. This was shown as an intermediate step in the proof of Theorem 2.1 in [Ma:85]:

Lemma (A.4). *Let $f : \mathbb{R}^{m \times n} \to \mathbb{R}$ be a separately convex function satisfying for some $\beta > 0$*

$$0 \leq f(X) \leq \beta(|X|^p + 1)$$

for all $X \in \mathbb{R}^{m \times n}$. Then there exists a constant $C > 0$ (depending only on β, p, m and n) such that

$$|f(X) - f(Y)| \leq C(1 + |X|^{p-1} + |Y|^{p-1})|X - Y|$$

for all $X, Y \in \mathbb{R}^{m \times n}$.

In Section 8.3 we employ a local version without the assumption on the upper bound from Lemma 2.2 in [BKK:00]:

Lemma (A.5). *If $f : B_{2r}(X_0) \to \mathbb{R}$ is a separately convex, then*

$$\text{lip}(f; B_r(X_0)) \leq \sqrt{mn} \frac{\text{osc}(f; B_{2r}(X_0))}{r}.$$

A function $f : U \to \mathbb{R}$ is said to be separately convex if its extension by ∞ is a separately convex function $\mathbb{R}^{m \times n} \to \mathbb{R} \cup \{\infty\}$. (The condition in Definition (A.1) makes sense also if the function takes infinite values.) $\text{lip}(f; U)$ denotes the Lipschitz constant of f on U and

$$\text{osc}(f; U) = \sup\{|f(X) - f(Y)| : X, Y \in U\}$$

its oscillation.

Appendix B

Γ-convergence

Here we recall the definition and some important properties of Γ-convergence that are used frequently in our proofs. The not referenced definitions and claims are taken from Chapter 1 of [Br:02].

Definition (B.1). Let $\mathcal{F}_j : M \to [-\infty, \infty]$, $j \in \mathbb{N}$, be a sequence of functionals on a metric space (M, d). For $x \in M$ we define the Γ-lim inf and Γ-lim sup at x as

$$\Gamma(d)\text{-}\liminf_{j \to \infty} \mathcal{F}_j(x) \quad := \quad \inf \Big\{ \liminf_{j \to \infty} \mathcal{F}_j(x_j) : x_j \to x \Big\},$$

$$\Gamma(d)\text{-}\limsup_{j \to \infty} \mathcal{F}_j(x) \quad := \quad \inf \Big\{ \limsup_{j \to \infty} \mathcal{F}_j(x_j) : x_j \to x \Big\}.$$

If these values are equal, then we call it the Γ-*limit* of $\{\mathcal{F}_j\}_{j \in \mathbb{N}}$ in x and write

$$\Gamma(d)\text{-}\lim_{j \to \infty} \mathcal{F}_j(x).$$

Equivalently, $\mathcal{F}_\infty(x) = \Gamma(d)\text{-}\lim_{j \to \infty} \mathcal{F}_j(x)$ if the following conditions are satisfied:

- (lim inf-inequality) If $x_j \to x$ in M, then $\liminf_{j \to \infty} \mathcal{F}_j(x_j) \geq \mathcal{F}_\infty(x)$.

- (recovery sequence) There exists a sequence $x_j \to x$ in M such that $\lim_{j \to \infty} \mathcal{F}_j(x_j) = \mathcal{F}_\infty(x)$.

We say that $\{\mathcal{F}_j\}_{j \in \mathbb{N}}$ Γ-*converges* to some functional $\mathcal{F}_\infty$, if and only if it Γ-converges to $\mathcal{F}_\infty(x)$ at every $x \in M$.

The following theorem shows the importance of the Γ-convergence in the calculus of variations.

Theorem (B.2). *Let $\mathcal{F}_j : M \to (-\infty, \infty]$, $j \in \mathbb{N}$, be a sequence of functionals. Suppose*

- *there exists a compact set $K \subset M$ with $\inf_{x \in K} \mathcal{F}_j(x) = \inf_{x \in M} \mathcal{F}_j(x)$ for all $j \in \mathbb{N}$,*

- $\Gamma(d)\text{-}\lim_{j \to \infty} \mathcal{F}_j = \mathcal{F}_\infty.$

Then

$$\exists \min_{x \in M} \mathcal{F}_\infty(x) = \lim_{j \to \infty} \inf_{x \in M} \mathcal{F}_j(x).$$

Moreover, if $\{x_j\}_{j \in \mathbb{N}}$ is a precompact sequence such that $\lim_{j \to \infty} \mathcal{F}_j(x_j) = \lim_{j \to \infty} \inf_{x \in M} \mathcal{F}_j(x)$, then every limit of a subsequence of $\{x_j\}_{j \in \mathbb{N}}$ is a minimum point for $\mathcal{F}_\infty$.

Γ-limits are always lower semicontinuous. In point of fact, even (pointwisely defined) $\Gamma(d)\text{-}\liminf_{j\to\infty}\mathcal{F}_j$ and $\Gamma(d)\text{-}\limsup_{j\to\infty}\mathcal{F}_j$ are always lower semicontinuous.

Γ-convergence possesses the Urysohn property:

Theorem (B.3). *Take* $\lambda \in [-\infty,\infty]$ *and* $x \in M$. *Then* $\lambda = \Gamma(d)\text{-}\lim_{j\to\infty}\mathcal{F}_j(x)$ *if and only if for every subsequence* $\{\mathcal{F}_{j_k}\}_{k\in\mathbb{N}}$ *there exists a further subsequence* $\{\mathcal{F}_{j_{k_l}}\}_{l\in\mathbb{N}}$ *such that* $\lambda = \Gamma(d)\text{-}\lim_{l\to\infty}\mathcal{F}_{j_{k_l}}(x)$.

In some cases we a priori know that a sequence Γ-converges:

- If we have a constant sequence, i.e., $\mathcal{F}_j = \mathcal{F}$ for all $j \in \mathbb{N}$, then

$$\Gamma(d)\text{-}\lim_{j\to\infty}\mathcal{F}_j = \operatorname{lsc}\mathcal{F}$$

  where lsc stands for lower semicontinuous envelope (in the metric d).

- For a non-increasing sequence $\{\mathcal{F}_j\}_{j\in\mathbb{N}}$, it holds

$$\Gamma(d)\text{-}\lim_{j\to\infty}\mathcal{F}_j = \operatorname{lsc}\left(\lim_{j\to\infty}\mathcal{F}_j\right) = \operatorname{lsc}\left(\inf_{j\in\mathbb{N}}\mathcal{F}_j\right).$$

- If $\{\mathcal{F}_j\}_{j\in\mathbb{N}}$ is non-decreasing, then

$$\Gamma(d)\text{-}\lim_{j\to\infty}\mathcal{F}_j = \lim_{j\to\infty}\left(\operatorname{lsc}\mathcal{F}_j\right) = \sup_{j\in\mathbb{N}}\left(\operatorname{lsc}\mathcal{F}_j\right).$$

- $\{\mathcal{F}_j\}_{j\in\mathbb{N}}$ Γ-converges if and only if $\{\operatorname{lsc}\mathcal{F}_j\}_{j\in\mathbb{N}}$ Γ-converges since

$$\begin{aligned}\Gamma(d)\text{-}\liminf_{j\to\infty}\mathcal{F}_j &= \Gamma(d)\text{-}\liminf_{j\to\infty}\left(\operatorname{lsc}\mathcal{F}_j\right),\\ \Gamma(d)\text{-}\limsup_{j\to\infty}\mathcal{F}_j &= \Gamma(d)\text{-}\limsup_{j\to\infty}\left(\operatorname{lsc}\mathcal{F}_j\right).\end{aligned}$$

- On a separable metric space every sequence of functionals always contains at least a subsequence that Γ-converges.

For a suitable class of integral functionals the following stronger compactness result holds (see, e.g., Theorem 12.5 in [BD:98]):

Theorem (B.4). *Let* $f_\varepsilon : \Omega \times \mathbb{R}^{m\times n} \to \mathbb{R}$, $\varepsilon > 0$, *be a family of Borel functions which for some* $\alpha, \beta > 0$ *satisfy the estimate*

$$\alpha|X|^p - \beta \le f_\varepsilon(x,X) \le \beta(1 + |X|^p)$$

for all $x \in \Omega$ *and* $X \in \mathbb{R}^{m\times n}$. *Define for* $U \in \mathcal{A}(\Omega)$ *and* $u \in W^{1,p}(\Omega;\mathbb{R}^m)$

$$\mathcal{F}_\varepsilon(u,U) := \int_U f_\varepsilon(x,\nabla u(x))\ dx,$$

and extend the definition to $L^p(\Omega;\mathbb{R}^m)$ *by* ∞. *Then, for every subsequence* $\varepsilon_j \searrow 0$ *there exists a further subsequence* $\{\varepsilon_{j_k}\}_{k\in\mathbb{N}}$ *and a Carathéodory function* $\phi : \Omega \times \mathbb{R}^{m\times n} \to \mathbb{R}$ *satisfying the same growth estimate as* $\{f_\varepsilon\}_\varepsilon$ *such that*

$$\mathcal{F}_0(u,U) := \int_U \phi(x,\nabla u(x))\ dx = \Gamma(L^p)\text{-}\lim_{k\to\infty}\mathcal{F}_{\varepsilon_{j_k}}(u,U)$$

for all $u \in W^{1,p}(\Omega;\mathbb{R}^m)$ *and* $U \in \mathcal{A}(\Omega)$.

A related constant sequence Γ-converges towards the lower semicontinuous envelope. In its calculation, Statement III.7 from [AF:86] plays a crucial role:

Theorem (B.5). *Let $\Omega \subset \mathbb{R}^n$ be a bounded open set. Suppose $f : \Omega \times \mathbb{R}^{m \times n} \to \mathbb{R}$ is a Carathéodory function which satisfies for some $p \geq 1$ and $\beta > 0$*

$$0 \leq f(x, X) \leq \beta(1 + |X|^p)$$

for almost all $x \in \Omega$ and all $X \in \mathbb{R}^{m \times n}$. Define for $u \in L^p(\Omega; \mathbb{R}^m)$

$$\mathcal{F}(u) := \begin{cases} \int_\Omega f(x, \nabla u(x)) \ dx, & u \in W^{1,p}(\Omega; \mathbb{R}^m), \\ \infty, & else. \end{cases}$$

The sequentially weakly lower semicontinuous envelope of $\mathcal{F}|_{W^{1,p}(\Omega; \mathbb{R}^m)}$ is given by

$$\mathrm{swlsc}\, \mathcal{F}|_{W^{1,p}(\Omega; \mathbb{R}^m)}(u) = \int_\Omega f^{\mathrm{qc}}(x, \nabla u(x)) \ dx$$

for each $u \in W^{1,p}(\Omega; \mathbb{R}^m)$.

In particular, if $p > 1$ and additionally $f(x, X) \geq \alpha |X|^p - \beta$, then the L^p-lower semicontinuous envelope of $\mathcal{F}$ is $\mathrm{swlsc}\, \mathcal{F}|_{W^{1,p}(\Omega; \mathbb{R}^m)}$ extended by ∞ to $L^p(\Omega)$.

Appendix C

Miscellaneous results

C.1 Besicovitch derivation theorem

In the analysis of measures, the Besicovitch derivation theorem plays a major role. We give a version from Theorem 1.153 in [FL:07].

Theorem (C.1). *Let μ, ν be two positive regular Borel measures on $\mathbb{R}^n$. There exists a Borel set $N \subset \mathbb{R}^n$ with $\mu(N) = 0$ such that for any $x \in \mathbb{R}^n \setminus N$ and any convex compact neighbourhood of the origin $C \subset \mathbb{R}^n$*

$$\frac{d\nu^a}{d\mu}(x) = \lim_{r \searrow 0} \frac{\nu(x + rC)}{\mu(x + rC)} \in \mathbb{R}$$

and

$$\lim_{r \searrow 0} \frac{\nu^s(x + rC)}{\mu(x + rC)} = 0,$$

where

$$\nu = \nu^a + \nu^s, \quad \nu^a \ll \mu \quad and \quad \nu^s \perp \mu.$$

C.2 Weakly convergent measures

The following results for Borel measures on an open subset $\Omega \subset \mathbb{R}^n$ that we apply in Chapter 8 may be found, e.g., in [ABM:06] under Lemma 4.2.1 and Corollary 4.2.1.

Lemma (C.2). *Let μ be a finite positive Borel measure on Ω and $\{B_i : i \in I\}$ a family of pairwise disjoint Borel subsets of Ω. Then the set*

$$\{i \in I : \mu(B_i) \neq 0\}$$

is at most countable.

Lemma (C.3). *Suppose*

$$\mu_j \xrightarrow{*} \mu \quad in \ M(\Omega; \mathbb{R}^N) \quad and \quad |\mu_j| \xrightarrow{*} \sigma \quad in \ M(\Omega).$$

Then

$$\lim_{j \to \infty} \mu_j(B) = \mu(B)$$

for every relatively compact Borel subset $B \subset \Omega$ with $\sigma(\partial B) = 0$.

C.3 Equiintegrable modifications

The following equiintegrability result of Fonseca, Müller and Pedregal is crucial in the proof of our main closure theorem (2.2), see Lemma 1.2 in [FMP:98] and Lemma 8.3 in [Pe:97]:

Lemma (C.4). *Let $\Omega \subset \mathbb{R}^n$ be an open bounded set, and let $\{u_i\}_{i\in\mathbb{N}}$ be a bounded sequence in $W^{1,p}(\Omega; \mathbb{R}^m)$, $1 < p < \infty$. There exist a subsequence $\{u_{i_k}\}_{k\in\mathbb{N}}$ and a sequence $\{v_k\}_{k\in\mathbb{N}} \subset W^{1,p}(\Omega; \mathbb{R}^m)$ such that*

$$\lim_{k\to\infty} \left| \{\nabla v_k \neq \nabla u_{i_k}\} \cup \{v_k \neq u_{i_k}\} \right| = 0$$

and $\{|\nabla v_k|^p\}_{k\in\mathbb{N}}$ is equiintegrable. Moreover, if $u_i \rightharpoonup u$ in $W^{1,p}(\Omega; \mathbb{R}^m)$, then the v_k can be chosen in such a way that $v_k = u$ on $\partial\Omega$ and $v_k \rightharpoonup u$ in $W^{1,p}(\Omega; \mathbb{R}^m)$.

C.4 Geometric rigidity

The following geometric rigidity theorem, proved in [FJM:02] and extended to general p in [CS:06], is a key step in the application of our abstract results to elasticity theory.

Theorem (C.5). *Suppose that $\Omega \subset \mathbb{R}^n$ is a bounded domain with Lipschitz boundary and $1 < p < \infty$. Then there exists a constant $C > 0$ such that for all $u \in W^{1,p}(\Omega; \mathbb{R}^n)$ there is an $R \in SO(n)$ with*

$$\|\nabla u - R\|_{L^p(\Omega; \mathbb{R}^{n\times n})} \leq C \| \operatorname{dist}(\nabla u, SO(n)) \|_{L^p(\Omega)}.$$

This theorem can be seen as a nonlinear variant of Korn's inequality, where instead of rotations R and the distance from $SO(n)$, one has an analogous estimate for the distance to the set $\mathbb{R}^{n\times n}_{\mathrm{skw}}$ of skew-symmetric matrices in terms of a single matrix in $\mathbb{R}^{n\times n}_{\mathrm{skw}}$. Below we give Korn's inequality in the so-called general case.

Theorem (C.6). *Suppose that $\Omega \subset \mathbb{R}^n$ is a bounded domain with Lipschitz boundary and $1 < p < \infty$. Then there exists a constants $C > 0$ such that for all $u \in W^{1,p}(\Omega; \mathbb{R}^n)$*

$$\|\nabla u\|_{L^p(\Omega; \mathbb{R}^{n\times n})} \leq C \big(\|\mathfrak{E}u\|_{L^p(\Omega; \mathbb{R}^{n\times n})} + \|u\|_{L^p(\Omega; \mathbb{R}^n)} \big).$$

Bibliography

[AF:86] E. ACERBI, F. FUSCO. *Semicontinuity problems in the Calculus of variations.* Arch. Ration. Mech. Anal. **86** (1984), 125–145.

[AK:81] M. AKCOGLU, U. KRENGEL. *Ergodic theorems for superadditive processes.* J. Reine Angew. Math. **323** (1981), 53–67.

[Al:93] G. ALBERTI. *Rank one property for derivatives of functions with bounded variation.* Proc. Roy. Soc. Edinburgh Sect. A **123** (1993), 239–274.

[ABC:14] G. ALBERTI, S. BIANCHINI, G. CRIPPA. *On the L^p-differentiability of certain classes of functions.* Rev. Mat. Iberoam. **30** (2014), no. 1, 349–367.

[AB:97] J. ALIBERT, G. BOUCHITTÉ. *Non-uniform integrability and generalized Young measures.* J. Convex Anal. **4** (1997), no. 1, 129–147.

[AM:02] F. ALVAREZ, J. MANDALLENA. *Homogenization of multiparameter integrals.* Nonlinear Anal. **50** (2002), no. 6, Ser. A: Theory Methods, 839–870.

[AM:04] F. ALVAREZ, J. MANDALLENA. *Multi-parameter homogenization by localization and blow-up.* Proc. Roy. Soc. Edinburgh Sect. A **134** (2004), no. 5, 801–814.

[ACD:97] L. AMBROSIO, A. COSCIA, G. DAL MASO. *Fine properties of functions with bounded deformation.* Arch. Rational Mech. Anal. **139** (1997), no. 3, 201–238.

[AD:92] L. AMBROSIO, G. DAL MASO. *On the relaxation in $BV(\Omega; \mathbb{R}^m)$ of quasi-convex integrals.* J. Funct. Anal. **109** (1992), no. 1, 76–97.

[AFP:00] L. AMBROSIO, N. FUSCO, D. PALLARA. Functions of bounded variation and free discontinuity problems. The Clarendon Press, Oxford University Press, New York, 2000.

[AG:80] G. ANZELLOTTI, M. GIAQUINTA. *Existence of the displacement field for an elastoplastic body subject to Hencky's law and von Mises yield condition.* Manuscripta Math. **32** (1980), no. 1–2, 101–136.

[AG:82] G. ANZELLOTTI, M. GIAQUINTA. *On the existence of the fields of stresses and displacements for an elasto-perfectly plastic body in static equilibrium.* J. Math. Pures Appl. **9** 61 (1982), no. 3, 219–244 (1983).

[ABM:06] H. ATTOUCH, G. BUTTAZZO, G. MICHAILLE. Variational analysis in Sobolev and BV spaces. Applications to PDEs and optimization. MPS/SIAM Ser. Optim., Philadelphia, 2006.

141

[Ba:15] J. BABADJIAN. *Traces of functions of bounded deformation.* Indiana Univ. Math. J. **64** (2015), no. 4, 1271–1290.

[BF:07] M. BAÍA, I. FONSECA. *The limit behavior of a family of variational multiscale problems.* Indiana Univ. Math. J. **56** (2007), no. 1, 1–50.

[BKK:00] J. BALL, B. KIRCHHEIM, J. KRISTENSEN. *Regularity of quasiconvex envelopes.* Calc. Var. Partial Differential Equations **11** (2000), no. 4, 333–359.

[BFT:00] A. BARROSO, I. FONSECA, R. TOADER. *A relaxation theorem in the space of functions of bounded deformation.* Ann. Scuola Norm. Sup. Pisa Cl. Sci. (4) **29** (2000), no. 1, 19–49.

[Bh:06] K. BHATTACHARYA. Microstructure of martensite. Why it forms and how it gives rise to the shape-memory effect. Oxford University Press, Oxford, 2003.

[BS:90] W. BORCHERS, H. SOHR. *On the equations* rot $v = g$ *and* div $u = f$ *with zero boundary conditions.* Hokkaido Math. J. **19** (1990), no. 1, 67–87.

[BD:93] G. BOUCHITTÉ, G. DAL MASO. *Integral representation and relaxation of convex local functionals on* BV(Ω). Ann. Scuola Norm. Sup. Pisa Cl. Sci. (4) **20** (1993), no. 4, 483–533.

[BFM:98] G. BOUCHITTÉ, I. FONSECA, L. MASCARENHAS. *A global method for relaxation.* Arch. Rational Mech. Anal. **145** (1998), no. 1, 51–98.

[Br:85] A. BRAIDES. *Homogenization of some almost periodic functionals.* Rend. Accad. Naz. Sci. XL **103** (1985), 313–322.

[Br:86] A. BRAIDES. *A homogenization theorem for weakly almost periodic functionals.* Rend. Accad. Naz. Sci. XL **104** (1986), 261–281.

[Br:92] A. BRAIDES. *Almost periodic methods in the theory of homogenization.* Appl. Anal. **47** (1992), no. 4, 259–277.

[Br:02] A. BRAIDES. Γ-convergence for beginners. Oxford University Press, Oxford, 2002.

[BD:98] A. BRAIDES, A. DEFRANCESCHI. Homogenization of multiple integrals. The Clarendon Press, Oxford University Press, New York, 1998.

[BDV:97] A. BRAIDES, A. DEFRANCESCHI, E. VITALI. *A relaxation approach to Hencky's plasticity.* Appl. Math. Optim. **35** (1997), no. 1, 45–68.

[Bre:13] K. BREDIES. *Symmetric tensor fields of bounded deformation.* Ann. Mat. Pura Appl. (4) **192** (2013), no. 5, 815–851.

[BF:91] G. BUTTAZZO, L. FREDDI. *Functionals defined on measures and applications to non-equi-uniformly elliptic problems.* Ann. Mat. Pura Appl. (4) **159** (1991), 133–149.

[CS:06] S. CONTI, B. SCHWEIZER. *Rigidity and Gamma convergence for solid-solid phase transitions with* SO(2) *invariance.* Comm. Pure Appl. Math. **59** (2006), no. 6, 830–868.

[Dac:08] B. Dacorogna. Direct methods in the calculus of variations. Second edition. Springer, New York, 2008.

[Dai:06] S. Dain. *Generalized Korn's inequality and conformal Killing vectors.* Calc. Var. Partial Differential Equations **25** (2006), no. 4, 535–540.

[DM:86-1] G. Dal Maso, L. Modica. *Nonlinear stochastic homogenization.* Ann. Mat. Pura Appl. (4) **144** (1986), 347–389.

[DM:86-2] G. Dal Maso, L. Modica. *Nonlinear stochastic homogenization and ergodic theory.* J. Reine Angew. Math. **368** (1986), 28–42.

[DNP:02] G. Dal Maso, M. Negri, D. Percivale. *Linearized elasticity as Γ-limit of finite elasticity.* Calculus of variations, nonsmooth analysis and related topics. Set-Valued Anal. **10** (2002), no. 2–3, 165–183.

[DG:75] E. De Giorgi. *Sulla convergenza di alcune successioni d'integrali del tipo dell'area.* Rend. Mat. (6) **8** (1975), 277–294.

[DT:84] F. Demengel, R. Temam. *Convex functions of a measure and applications.* Indiana Univ. Math. J. **33** (1984), no. 5, 673–709.

[DT:86] F. Demengel, R. Temam. *Convex function of a measure: the unbounded case.* In FERMAT days 85: mathematics for optimization. Ed. by J. Hiriart-Urruty. North-Holland Publishing Co., Amsterdam, 1986.

[DQ:90] F. Demengel, T. Qi. *Convex function of a measure obtained by homogenization.* SIAM J. Math. Anal. **21** (1990), no. 2, 409–435.

[DR:16] G. De Philippis, F. Rindler. *On the structure of 𝒜-free measures and applications.* Ann. of Math. (2) **184** (2016), no. 3, 1017–1039.

[DR:17] G. De Philippis, F. Rindler. *Characterization of generalized Young measures generated by symmetric gradients.* arXiv:1604.04097 [math.AP]

[DM:87] R. DiPerna, A. Majda. *Oscillations and concentrations in weak solutions of the incompressible fluid equations.* Comm. Math. Phys. **108** (1987), no. 4, 667–689.

[DG:16] M. Duerinckx, A. Gloria. *Stochastic homogenization of nonconvex unbounded integral functionals with convex growth.* Arch. Ration. Mech. Anal. **221** (2016), no. 3, 1511–1584.

[DL:76] G. Duvaut, J. Lions. Inequalities in mechanics and physics. Springer-Verlag, Berlin-New York, 1976.

[FL:07] I. Fonseca, G. Leoni. Modern methods in the calculus of variations: L^p spaces. Springer Monographs in Mathematics. Springer, New York, 2007.

[FMP:98] I. Fonseca, S. Müller, P. Pedregal. *Analysis of oscillation and concentration effects generated by gradients.* SIAM J. Math. Anal. **29** (1998), 736–756.

[FJM:02] G. Friesecke, R. James, S. Müller. *A theorem on geometric rigidity and the derivation of nonlinear plate theory from three-dimensional elasticity.* Comm. Pure Appl. Math. **55** (2002), no. 11, 1461–1506.

[FR:10] M. Fuchs, S. Repin. *Some Poincaré-type inequalities for functions of bounded deformation involving the deviatoric part of the symmetric gradient.* Zap. Nauchn. Sem. S.-Peterburg. Otdel. Mat. Inst. Steklov. (POMI) **385** (2010), 224–233.

[GT:01] D. Gilbarg, N. Trudinger. Elliptic partial differential equations of second order. Springer-Verlag, Berlin, 2001.

[GN:11] A. Gloria, S. Neukamm. *Commutability of homogenization and linearization at identity in finite elasticity and applications.* Ann. Inst. H. Poincaré Anal. Non Linéaire **28** (2011), no. 6, 941–964.

[Gr:90] G. Grubb. *Pseudo-differential boundary problems in L_p spaces.* Comm. Partial Differential Equations **15** (1990), no. 3, 289–340.

[Ha:96] P. Hajlasz. *On approximate differentiability of functions with bounded deformation.* Manuscripta Math. **91** (1996), no. 1, 61–72.

[HR:99] W. Han, B. Reddy. Plasticity. Mathematical theory and numerical analysis. Springer-Verlag, New York, 1999.

[HN:80] I. Hlávaček, J. Nečas. Mathematical theory of elastic and elasto-plastic bodies: an introduction. Elsevier Scientific Publishing Co., Amsterdam-New York, 1980.

[JS:14] M. Jesenko, B. Schmidt. *Closure and commutability results for Γ-limits and the geometric linearization and homogenization of multiwell energy functionals.* SIAM J. Math. Anal. **46** (2014), no. 4, 2525–2553.

[KK:16] B. Kirchheim, J. Kristensen. *On rank one convex functions that are homogeneous of degree one.* Arch. Ration. Mech. Anal. **221** (2016), no. 1, 527–558.

[Kr:94] J. Kristensen. Finite functionals and Young measures generated by gradients of Sobolev functions. Mat-report 1994-34. Mathematical Institute, Technical University of Denmark, 1994.

[KR:10-1] J. Kristensen, F. Rindler. *Relaxation of signed integral functionals in BV.* Calc. Var. Partial Differential Equations **37** (2010), no. 1–2, 29–62.

[KR:10-2] J. Kristensen, F. Rindler. *Characterization of generalized gradient Young measures generated by sequences in $W^{1,1}$ and BV.* Arch. Ration. Mech. Anal. **197** (2010), no. 2, 539–598.

[LM:02] C. Licht, G. Michaille. *Global-local subadditive ergodic theorems and application to homogenization in elasticity.* Ann. Math. Blaise Pascal **9** (2002), no. 1, 21–62

[Ma:78] P. MARCELLINI. *Periodic solutions and homogenization of nonlinear variational problems.* Ann. Mat. Pura Appl. (4) **117** (1978), 139–152.

[Ma:85] P. MARCELLINI. *Approximation of quasiconvex functions, and lower semicontinuity of multiple integrals.* Manuscripta Math. **51** (1985), no. 1–3, 1–28

[MM:94] K. MESSAOUDI, G. MICHAILLE. *Stochastic homogenization of nonconvex integral functionals.* RAIRO Modél. Math. Anal. Numér. **28** (1994), no. 3, 329–356.

[Mo:16] M. MORA. *Relaxation of the Hencky model in perfect plasticity.* J. Math. Pures Appl. (9) **106** (2016), no. 4, 725–743.

[Mü:87] S. MÜLLER. *Homogenization of nonconvex integral functionals and cellular elastic materials.* Arch. Ration. Mech. Anal. **99** (1987), no. 3, 189–212.

[MN:11] S. MÜLLER, S. NEUKAMM. *On the commutability of homogenization and linearization in finite elasticity.* Arch. Ration. Mech. Anal. **201** (2011), no. 2, 465–500.

[Ne:10] S. NEUKAMM. Homogenization, linearization and dimension reduction in elasticity with variational methods. Ph.D. thesis, Technische Universität München, 2010.

[Pe:97] P. PEDREGAL. Parametrized measures and variational principles. Birkhäuser Verlag, Basel, 1997.

[Re:94] Y. RESHETNYAK. Stability theorems in geometry and analysis. Kluwer Academic Publishers Group, Dordrecht, 1994.

[Sch:08] B. SCHMIDT. *Γ-limits of multiwell energies in nonlinear elasticity theory.* Contin. Mech. Thermodyn. **20** (2008), no. 6, 375–396.

[Su:81] P. SUQUET. *Sur les équations de la plasticité: existence et régularité des solutions.* J. Mécanique **20** (1981), no. 1, 3–39.

[Te:85] R. TEMAM. Mathematical problems in plasticity. Gauthier-Villars, 1985.

[WY:06] Z. WU, J. YIN, C. WANG. Elliptic & parabolic equations. World Scientific Publishing Co. Pte. Ltd., Hackensack, NJ, 2006.

[Ze:16] C. ZEPPIERI. *Stochastic homogenisation of singularly perturbed integral functionals.* Ann. Mat. Pura Appl. (4) **195** (2016), no. 6, 2183–2208.

[Zh:97] K. ZHANG. *Quasiconvex functions, SO(n) and two elastic wells.* Ann. Inst. H. Poincaré Anal. Non Linéaire **14** (1997), no. 6, 759–785.

[Zh:04] K. ZHANG. *An approximation theorem for sequences of linear strains and its applications.* ESAIM Control Optim. Calc. Var. **10** (2004), 224–242.

[Zi:89] W. ZIEMER. Weakly differentiable functions. Sobolev spaces and functions of bounded variation. Springer-Verlag, New York, 1989.